# Postphenomenological Investigations

# Postphenomenology and the Philosophy of Technology



Technological advances affect everything from our understandings of ethics, politics, and communication, to gender, science, and selfhood. Philosophical reflection on technology helps draw out and analyze the nature of these changes, and helps us understand both the broad patterns and the concrete details of technological effects. This book series provides a publication outlet for the field of the philosophy of technology in general, and the school of thought called "postphenomenology" in particular. Philosophy of technology applies insights from the history of philosophy to current issues in technology, and reflects on how technological developments change our understanding of philosophical issues. In response, postphenomenology analyzes human relationships with technologies, while integrating philosophical commitments of the American pragmatist tradition of thought.

*Design, Meditation, and the Posthuman*, Edited by Dennis M. Weiss, Amy D. Propen, and Colbey Emmerson Reid

*Postphenomenological Investigations: Essays on Human-Technology Relations*, Edited by Robert Rosenberger and Peter-Paul Verbeek.

# Postphenomenological Investigations

## *Essays on Human–Technology Relations*

Edited by Robert Rosenberger
and Peter-Paul Verbeek

LEXINGTON BOOKS
Lanham • Boulder • New York • London

Published by Lexington Books
An imprint of The Rowman & Littlefield Publishing Group, Inc.
4501 Forbes Boulevard, Suite 200, Lanham, Maryland 20706
www.rowman.com

Unit A, Whitacre Mews, 26-34 Stannary Street, London SE11 4AB



British Library Cataloguing in Publication Information Available

**Library of Congress Cataloging-in-Publication Data**

Postphenomenological investigations : essays on human-technology relations / edited by Robert Rosenberger and Peter-Paul Verbeek.
pages cm -- (Postphenomenology and the philosophy of technology)
Includes bibliographical references and index.
ISBN 978-0-7391-9436-2 (cloth : alk. paper) -- ISBN 978-0-7391-9437-9 (electronic) 1. Technology--Philosophy--History--20th century. 2. Phenomenology. 3. Postmodernism. I. Rosenberger, Robert (Robert Joseph), editor. II. Verbeek, Peter-Paul, 1970- editor.
T14.P576 2015
303.48'301--dc23
2015004561

∞™ The paper used in this publication meets the minimum requirements of American National Standard for Information Sciences Permanence of Paper for Printed Library Materials, ANSI/NISO Z39.48-1992.

Printed in the United States of America

# Table of Contents

# Preface

## *Positioning Postphenomenology*

Don Ihde

Postphenomenology is a philosophical style of analysis which deals with science and technology studies. It is a recent comer to a series of twentieth and twenty-first century interpretive upheavals regarding science, technology, cultural studies, and technoscience. But where does one begin? If there is a "standard view," then early modern science is usually thought to begin in the seventeenth century. Yet the term, "scientist," did not occur or begin to be used until the nineteenth century. William Whewell coined the term in 1833. Before this neologism, "natural philosophy" was the dominant term used.

What this shift in terms points to is a deep question: who shall interpret science? Early interpreters would have been identified, anachronistically, as scientists (Galileo, Leewenhoek) or as philosophers (Bacon, Descartes). But if, as I have claimed, early modern science takes much of its specific shape through its technologies—instruments, early often optical—then one must also account for science *and* technology. And here the same enigma occurs, since the term "technology" does not come into common usage until even later than "scientist." Historians Thomas Hughes and David Nye have pointed out that "technology" as a term emerged in the early twentieth century. Prior to technology, there were "industrial arts" and "engineering" or "machines." In short, new terminology arrives after, later than, the shifts which have already occurred concerning science and technology. By the time Whewell coins "scientist" the times already demonstrated how secular, desacralized modes of reasoning could be recognized. Similarly, only after massive industrialization could "technology" be recognized.

Here my issue relates to postphenomenology as a particular mode of science-technology interpretation. Its arrival coincides with a late-twentieth- to twenty-first-century radical shift in science-technology analysis. I shall take 1979 as a watershed year by highlighting two publications of that year: Bruno Latour and Steve Woolgar, *Laboratory Life: The [Social] Construction of Scientific Facts* and my *Technics and Praxis: A Philosophy of Technology.* Latour's book is usually taken as a landmark book in a new sociology of science. (It is interesting to note that the original subtitle was the "*social* construction" of scientific facts, but *social* was dropped from the subtitle in the second edition.) My book is frequently cited as the first English language book on philosophy of technology. (I note in passing that neither Latour nor I knew each other or knew of our respective books then, nor was "Actor-Network Theory" nor "postphenomenology" used as a method descriptor.)

I cannot take the time or space here to detail the several and important changes to the understanding of science from its early modern beginnings (seventeenth century) through its progression and growing autonomy (eighteenth century) to its elevation as the supreme mode of knowledge (nineteenth century). Rather, I shall concentrate upon a rapid set of changes beginning in the twentieth century. Early in the century dominant interpretations were centered in *mathematization* (Duhem, Mach, Poincare) followed by the rise of *positivism* (Vienna Circle, et al.) and the interpreters were scientists, philosophers, and historians—mostly mathematizers, logicians, and theory-oriented thinkers. Continuing the idealistic take on science as a linear, accumulative, and ahistorical-universal knowledge of nineteenth-century inheritance, this interpretation dominated. It began to be challenged by the mid-twentienth century with the rise of anti-positivism. Discontinuity theorists began to describe "scientific revolutions" and a science saturated with history (Kuhn, Foucault, Feyerabend, et al.). Science, beginning to be culturally and historically embedded, also began to be interpreted from a new institutionalization of interpretation through history and philosophy of science departments (1950s through the 1960s). Alongside the history and philosophy of science, there had also existed a sociology of science (Merton) which did look at science practice, but focused largely on outputs—such as citation research and rankings. This sociology of science chronologically coincided with the positivist–anti-positivist debates. However, by the 1970s, there began yet another more radical shift. Here the earliest origins were largely sociological. The United Kingdom was the early locale, with the *social studies of scientific knowledge* (SSK) and *social constructionism* as designators of the new approach (Bloor, Barnes, Collins). As with earlier Mertonian sociology, social constructive approaches continued to forefront science as practice, but unlike the earlier sociology, constructive approaches began to see that the content of scientific knowledge itself was "socially constructed."

It is at this point that phenomenology can be seen to play a background and tangential role. Berger and Luckmann's *The Social Construction of Reality* (1966) was inspired by the phenomenologies of Alfred Schutz and Edmund Husserl. And while phenomenology was muted in the Anglophone version of social constructionism, it retained the shift of interpretive focus on *praxis.* We now arrive at my chosen juncture of 1979 and the birth of what were to become Actor-Network Theory and postphenomenology.

In *Laboratory Life* Latour described himself as a sociologist and anthropologist. Laboratory culture was his substitute for a "tribe" and he would become a participant observer. So, as an observer, he would follow around the scientists and technicians and take account of their activities and outcomes. How is scientific knowledge produced? Here I confess, I came to this book late—years earlier I had read and taught his *Science in Action* (1987) which was a much more philosophical re-working of *Laboratory Life*, although most of the key ideas for the later book were already present in a more diaristic-interview form in the earlier. From the outset, it is obvious that Latour's approach is radically different from either the history and philosophy of science, or earlier sociologies of science. He self-consciously brings a "literary critical" approach to his descriptions of what scientists do. Although the index does not include the term, "semiotics," Latour's style of analysis was later dubbed a *material semiotics.* It should be noted, too, that his laboratory was that of Roger Guillemin in the Jonas Salk Institute and thus was a chemical-biological science. Thus, contrary to the sciences of interest to the history and philosophy of science traditions which often assumed the primacy of physics, mathematics (and astronomy); with Latour biology, chemistry, and neurology are analyzed.

I begin my account of *Laboratory Life* with what Latour depicts of the laboratory in his photograph file—it is a sort of geography of the lab. Three image themes stand out: machinery and instruments (five out of fourteen) are shown. People—sometimes also in the instrument images—are shown, usually in discussion or poring over reading material. And, reading material—print outs, scientific articles, and sometimes with machines which produce print. This is, of course, what the observer sees walking around the lab. Suppose now, imaginatively, a similar set of images first associated with early twentieth-century philosophy of science as mathematized, theory-producing—would the pictures be of an Einstein sitting and thinking a thought experiment? Of a mathematician doing chalk calculations on a blackboard? Or, imagine a Kuhnian image showing how an improved instrument yields a different paradigm shift which includes different modes of seeing. Kuhn points out that seventeenth-century effluvium theorists saw "chaff particles . . . fall off electrified bodies" (1970, 117). But later this phenomenon was seen as electrostatic repulsion, though it was "not seen as such until Hauksbee's large-scale apparatus had greatly magnified its effects" (Kuhn, 1970, 117). And again "Herschel, when he first observed the same object twelve years later . . . *with a much improved telescope* of his own manufacture . . . was able to notice an apparent disk-size that was at least unusual for the stars" (Kuhn, 1970, 115). What I am suggesting is that in early mathematized and theory-centered interpretations of science, instruments play no important role; later, the role of instruments (Kuhn) does get acknowledged, later accepted and then accelerated with Latour.

Moving now from Latour's three image themes to his sociology-anthropology—but also his hermeneutical-literary theory observations, I will now begin to sketch the new interpretation of science which results. I begin, not with Latour's opening perspective, but with the "technologies" of the lab. First, biochemical instruments differ from those in other sciences. Latour includes in his photos (1) gamma counter, (2) NMR spectrometer, (3) fractionating columns, and (4) an automatic amino acid analyzer. Most of these instruments relate to bioassays and the material which goes into the assay is biological, hormonal, chemical—including lots of rat brain and other animal material. Note that the object being constructed is TRF (or TRH) which requires "several tons of hypothalamic tissues obtained from the slaughterhouse" (Latour and Woolgar, 1986, 108). The instruments mentioned are used in the process of isolating and identifying TRF(H), the new scientific object of the study. Latour recognizes the important role played by instruments (technologies) but gives a unique description of this role: *instruments are inscription devices.* What the instruments produce is what later is described as a "visual display" but which here remains an *inscription.* This is a focal emphasis upon what Latour is drawing from the hermeneutic-literary dimension of his perspective. The instruments being used here—spectrometer and amino acid analyzer in particular—produce (visual) graphs. These are squiggle patterns common to a whole range of instruments (like an EKG) with peaks, spikes, and the like. In my terminology, these are non-isomorphic images which, unlike photographic, isomorphic images which resemble what is being imaged, are more text- or code-like patterns which are "read." Wave pattern oscilligraph-like printouts on rolling scroll paper are what one sees. One can see why the Derrida "inscription" or "trace" language works well here.[1]

I now move to the third theme of images—written material, scientific articles, lab notes, and the like. Latour as observer again takes a literary turn in his description of laboratory life—what do the humans produce? Texts, articles, notes. How these are developed, dealt with, and processed give clues to how scientific knowledge is constructed. I shall not detail all of the moves Latour makes, but briefly, only what survives becomes a scientific fact. And in the process—one could say stylistically—all traces of history, subjectivity, construction are removed. So, contrary to what he observes ethnographically, all the processing which goes on in the lab, gets stated in such a way that the object [TRF (H)] gets removed from its dynamic process. So, now we have instruments which produce inscriptions, which in turn are reported in published scientific articles claiming new scientific facts—but these are, in turn, described as having been discovered as though the object is purported to have always been there. In short, the rhetorical style of this literature eliminates the contingencies, battles, processes which occurred in the lab as observed by the sociologist-anthropologist.

We now need to look at the humans [the tribe] who operate the instruments and produce the writings. Here I turn to the concluding chapter, "The Creation of Order out of Disorder." Looking back at his observations, Latour describes a process by which the scientific object or fact (TRF) comes to be. Over all it only becomes stabilized after much processing—TRF is slowly *constructed* by going through many trials and manipulations, "after eight years of bringing inscription devices to bear on the purified brain extracts, the statement stabilized sufficiently to enable it to switch into another network" (Latour and Woolgar, 1986, 236). Now a reality, TRF is seen as a *consequence* of the process. Only by going through what Latour calls an *agonistic field,* can the object-fact (TR) be *reified* or *materialized.* One can see that in one sense, Latour has inverted the usual take on how scientific knowledge is produced. In Latour's perspective "Nature" is always the last to arrive, the *consequence* of the long construction, "Nature is a usable concept only as a byproduct of agonistic activity" (Latour and Woolgar, 1986, 327). The laboratory has been the site where construction takes place.

I may now turn to my own *Technics and Praxis: A Philosophy of Technology* which was also published in 1979. Philosophers rarely have had deep interests in technologies until late modernity. But after the Industrial Revolution, the World Wars, the impact of technologies could hardly be ignored. For philosophy, however, this would require a shift of temperament. The dominant temperament in philosophy, from Classical Greece on, favored the ideal, the abstract, and often shunned the material, or in the case of the long meditations on body and mind, ranked materiality as inferior. Thus with respect to science and technology, technology was almost always secondary, epiphenomenal, or "applied." I began to realize the importance of technologies, tools, materiality while still at Southern Illinois University. I had been assigned to do interdisciplinary work in the SIU Honors College where, in 1964, the theme was "the leisure society." Readings argued that modern technologies freed up time such that in modernity we should be able to create a "New Athens." I was skeptical of the leisure claim and began to reflect on *a phenomenology of work.* What stood out was how deeply virtually anything to do with work entailed some kind of tool, artifact, *technology.* I asked my students to do a sort of diary of their ordinary day and specifically account for relations with technologies. (Awaken with an alarm clock; engage a bathroom set of technologies; prepare breakfast with a toaster; in those days take notes with a pencil and pad [today laptops and iPads]; and so each would find multiple—hundreds—of such engagements in any given day.) At that time, too, I was reading seriously Heidegger on the primacy of *pragmata* over scientific objects in the human experience of World. He was probably the first to claim that technology is "ontologically prior" to science. By the 1970s, I was doing articles on human-machine relations with a special interest in the role of instruments in science. It was this work which became the first four chapters of *Technics and Praxis*—a set of phenomenological examinations of the different types [of human-technology relations including examples from science uses].

In this analysis I was asserting several new- and non-classical phenomenological modifications: I continued to use what I would later term, the *interrelational ontology* implied by Husserl's "intentionality" and Heidegger's "being-in-the-world," but into a mediating role I placed material technologies. Human-technology-World became the formalism expressing this interrelationality. Humans actionally using technologies mediatingly relate to a World. What was to become known as my "phenomenology of technics" began here with a descriptive analysis of *embodiment, hermeneutic,* and *background relations* [later to add *alterity relations* in *Technology and the Lifeworld,* 1990.] In *Technics and Praxis* these four chapters were called a "Program in Philosophy of Technology." This was clearly a praxis-oriented analysis, although not sociological. It was instead phenomenological with an emphasis upon how scientists and others were bodily-perceptually engaging a world through instruments. I shall not here go into detail since there are multiple references to this technics in this volume. But I did analyze the ways in which by using technologies meditatively one could see how these uses *non-neutrally* [my neologism] transformed experience. In my telescope example—to be repeated and refined over the years—an optically transformed celestial world is very different from an eyeball world. Indeed, through optics the Copernican version of the heavens finally overcomes the Ptolemaic world. The phenomenology of *Technics and Praxis* emphasized the role of *embodiment* in relation to technology uses. Unlike *Laboratory Life* phenomenological embodiment was not linguistically or textually focused. This difference arises because of the differences between a semiotic and a phenomenological background perspective. It remains a major difference even today.

I have now crossed the 1979 divide, but the chronological parallels continue—still independently—for several years. John Law, active as a frequent visitor and collaborator at the *Centre de Sociologie de l'innovation* and *the Ecole Nationale superiere des Mines de Paris* where Latour and his colleague Michel Callon were staffed, noted that Callon begin to use the term

"Actor-Network Theory" in about 1982 (Law, 2007). In my case, after much reading and discussion of the Richard Rorty inspired non-foundational pragmatism in the North American context, I gave lectures at Goteborg, Sweden, on a "non-foundational phenomenology." Although I had not yet coined "postphenomenology" virtually all interpreters see its outline in the 1986 publication of that earlier title (Goteborg). Similarly, both Latour and I published more philosophically refined versions of the 1979 books—Latour, *Science in Action* (1987) and me, *Technology and the Lifeworld* (1990). It is my opinion that both these later books are better and more coherent than their 1979 predecessors. By this time our paths begin to cross at conferences and readings (4/S, the Society for the Social Studies of Science, was the main organization for STS papers—I note that my own first attendance and presentation was in 1979, but I attended only sporadically until the 1990s). I was aware of *Science in Action* and through it, Latour played a significant role in my 1991 *Instrumental Realism: The Interface between Philosophy of Science and Philosophy of Technology.*

A side glance at the new interpretations of science and technology in the 1980s shows a rapidly burgeoning publication field: philosophy of technology finally took off with Albert Borgmann, Langdon Winner, Hubert Dreyfus, and Andrew Feenberg joining in with published books; feminist thinkers Donna Haraway, Sandra Harding, and Evelyn Fox-Keller all dealt with science-technology topics. Both Anglophone and Francophone sociologies (SSK and ANT) proliferated. There were also significant commonalities traversing this set of thinkers. All agreed that science, unlike its older transcendental and ahistorical interpretation, was culture-, history-, and even gender-situated. All these thinkers employed a version of an inter-relational framework interpreting how science and technology developed. We now reach the decade of the 90s. I shall now narrow the focus more specifically upon the emergence of *postphenomenology* (1993 was the publication year for my *Postphenomenology: Essays in the Postmodern Context*).

It was during the 1990s that I met Peter-Paul Verbeek, first as a PhD student doing interesting work on philosophy of technology in his dissertation. In 1997 Hans Achterhuis, Verbeek's older colleague, published *Van stoommachine tot cyborg: Denken over techniek in de nieuwe wereld,* later translated into English as *American Philosophy of Technology: The Empirical Turn* (2001). Verbeek's chapter, "Don Ihde: The Technological Lifeworld" was an insightful analysis of my work vis-à-vis Heidegger.[2] Verbeek was already deeply into comparing Actor-Network Theory to postphenomenology and during a 1998 lecture trip to the Netherlands, he asked me to give a seminar comparing these two approaches. Luckily, I had been teaching *Science in Action* in my philosophy of science classes so was able to oblige. More significantly, in the background, both of us had already begun to have access to graduate students who were attracted to these modes of science-technology studies.

In my case, I had long dreamed of having some kind of institute which would do such analyses, and by 1994, there began what was to be a long progression of Visiting Scholars, faculty, and post-docs but also dissertation students from both Europe and Asia. Monique Riphagen from the Netherlands and Sung Dong Kim from Korea with the project of translating *Technics and Praxis* into Korean (published, 1998); these were the first. We formed an informal reading group which continued until 1998 when the *technoscience research seminar* was made a permanent part of our Stony Brook doctoral program. We read widely, and early works included the famous "chicken debate" between proponents of SSK and ANT and our early roastees were often associated with these strands of thought. This continued until my retirement in 2012 and hosted many Visiting Scholars, producing both MAs and PhDs from Stony Brook and internationally. Stony Brook PhDs with technoscience dissertations in this period included Evan Selinger (Rochester Institute of Technology), Ken Yip (New York Tech), Robert Rosenberger (Georgia Institute of Technology), Kyle Whyte (Michigan State University) and Adam Rosenfeld (North Carolina A and M). Several of these still-junior philosophers are already name visible in the field, and most have presented in the various postphenomenology research panels discussed below. Additionally, in the early days three Danes and one Argentinian completed MAs, and visitors doing PhDs in their own universities (with me as advisor) included two Danish students, two from Norway, and one each from Sweden, Greece, the UK, and the Netherlands.

We called this operation the Technoscience Research Group with its technoscience research seminar. Its first published book was *Chasing Technoscience* (2003) edited by Evan Selinger and me. It dealt with materiality in the works of Donna Haraway, Don Ihde, Bruno Latour, and Andrew Pickering, with contributions by the principals and critical responses by six of the TRG participants. We also developed the tradition of "roasts." Each semester had a different theme, and a roastee was chosen—first to be read and discussed for some six weeks, then to be given a full three-hour roast. We ended up with philosophers of science (Joseph Rouse and Peter Galison), philosophers of technology (Albert Borgmann, Andrew Feenberg, Hubert Dreyfus, Langdon Winner, and Peter-Paul Verbeek) sociologists of science (Andrew Pickering, Harry Collins, Trevor Pinch) feminist thinkers (Donna Haraway, Sandra Harding, Evelyn Fox-Keller) a historian of science (Paul Forman) and ending with me as final roastee during my retirement year.

The seminar encouraged participants to formulate conference presentations for science-technology associations. While at first these were simply individual contributions during the 1990s and into the early 2000s, by 2006 we began to propose distinctly postphenomenology research panels to the

Society for Phenomenology and the Human Sciences, and added, in 2007, the Social Studies of Science, and the Society for Philosophy and Technology. These rapidly grew into multiple panels and today often are quadruple or quintuple. During this same time period, and parallel to the Stony Brook operation, Peter-Paul Verbeek had been directing a consortium MA program and had a series of PhD students in the Netherlands. Thus, participants from both Stony Brook and Twente plus other Asian, European, and North American universities began to be a recognizable group, today over one hundred participants.

While the list is too long for space here, I would point to a growing number of publications—books, special issues of journals, individual articles in journals—discussing postphenomenology. Verbeek also continued his interest in comparing ANT and Postphenomenology and formed a special panel at the Rotterdam 4/S (2008)—it was jammed, and some had to listen from the hallway. This theme was later taken up in a conference in Manchester UK (2009) and a 4/S conference in Arlington, Virginia (2010) Verbeek, too, published two widely cited books, *What Things Do* (Penn State Press, 2005) and *Moralizing Technology* (Chicago, 2011). And this volume joins the first series on postphenomenology, Postphenomenology and the Philosophy of Technology, published by Lexington, which begins a longer list of works arising out of postphenomenological research programs.

In conclusion I will briefly note some of the major comparative features of the two styles of analysis:

- Both ANT and Postphenomenology employ inter-relational ontologies. But the philosophical traditions out of which these come differ. ANT draws from semiotics of which the base is linguistic-textual. Postphenomenology draws from an embodiment analysis of human action and perception. Yet inter-relationality in either form sees that when something in a World changes, so does what is human change.
- Both ANT and Postphenomenology are materially sensitive. In ANT sensitivity to materiality took the shape of its distinctive human and non-human actants. ANT also holds to a general and equalized notion of symmetry between the humans and non-humans. As noted above, Law terms this a *material semiotics.* Postphenomenology does not hold to a strict symmetry, but to a series of gradations between types of activity. But it, too, is materially sensitive—for example, postphenomenology recognizes *instrumental intentionalities* or built-in selectivities in technologies. I am developing a *material hermeneutics* modeled upon the use of contemporary science instruments to "let things speak."
- Both methodologies have abandoned early modern "subject-object" dichotomies and substituted multiple transition and structure notions such as "translation" in ANT and "multistability" in postphenomenology.

- Postphenomenology retains a version of *variational theory* from phenomenology, which is a method to possibilize multiple structures in such a way that what is variant and what is invariant can be shown.

Currently, in STS circles on the Continent, it probably remains the case that ANT is the dominantly favored approach. I would note that at the Manchester conference there was a bit of the division of the house with the sociologists siding with ANT but the anthropologists more likely to be favorable to postphenomenology. Philosophy of technology is highly positive to postphenomenology and at the 2013 Lisbon meeting of SPT [Society for Philosophy and Technology] the phenomenology-postphenomenology and technology track had 34 presenters, the largest single methological group. Clearly the two styles of analysis are more complementary than combative.

## NOTES

1. See note 24 (Latour and Woolgar, 1979, 261). Latour acknowledges his debt to Derrida's terminology.
2. See Peter-Paul Verbeek's entry in (Achterhuis, 2001, 119-146).

## REFERENCES

Achterhuis, H. (2001). *American Philosophy of Technology: The Empirical Turn,* translated by R. Crease. Bloomington: Indiana University Press.

Berger, P., and T. Luckmann. (1966). *The Social Construction of Reality.* New York: Anchor Books.

Ihde, D. (1979). *Technics and Praxis: A Philosophy of Technology,* Boston Series in the Philosophy of Science. Dordrecht: Reidel Press.

Ihde, D. (1986). *On Non-Foundational Phenomenology.* Goteborg: Institionen for pedagogic, Fenomenografiska notiser 3.

Ihde, D. (1990). *Technology and the Lifeworld; From Garden to Earth.* Bloomington: Indiana University Press.

Ihde, D. (1991). *Instrumental Realism: The Interface between Philosophy of Science and Philosophy of Technology.* Bloomington: Indiana University Press.

Ihde, D. (1993). *Postphenomenology: Essays in the Postmodern Context.* Evanston: Northwestern University Press.

Kuhn, T. (1970). *The Structure of Scientific Revolutions, 2nd ediction.* Chicago: University of Chicago Press.

Latour, B. (1987). *Science in Action.* Cambridge: Harvard University Press.

Latour, B., and S. Woolgar. (1979). *Laboratory Life: The Construction of Scientific Facts, 2nd edition.* Princeton: Princeton University Press. (1st edition, 1979).

Law, J. (2007). "Actor Network Theory and Material Semiotics." ANT document, April 2007. http://www.heterogeneities.net/publications/Law2007ANTandMaterialSemiotics.pdf

Verbeek, P-P. (2005). *What Things Do.* University Park, PA: Penn State University Press.

Verbeek, P-P. (2011). *Moralizing Technology.* Chicago: University of Chicago Press.

# Introduction

Robert Rosenberger and Peter-Paul Verbeek

A philosophical perspective called postphenomenology has been quickly gaining influence in discussions on technology in the Philosophy of Technology, Science and Technology Studies, and other fields. This developing school of thought brings together an international group of scholars working within a number of disciplines, including anthropology, sociology, cultural studies, media studies, as well as philosophy. This book serves as both an introduction to the postphenomenological perspective, as well as a review of some of the work on the leading edge of this school of thought.

Postphenomenologists study the relationships that develop between users and technologies. This perspective addresses questions such as: How do technologies shape our choices, our actions, and our experience of the world? How are technologies at once objects that we use for our own purposes, and at the same time objects that have an influence on us? How do technologies inform our politics, ethics, and our understandings of the basic features of our everyday experience?

As a distinctive philosophical perspective, postphenomenology brings together a number of insights, interests, and commitments. It builds on the phenomenological tradition of philosophy, approaching issues from the starting point of deep descriptions of human experience. Postphenomenology brings together the phenomenological approach and the ontological commitments of the American pragmatist tradition of philosophy. This means that postphenomenological claims are never about the absolute foundations of reality or knowledge, and never about the "essence" of an object of study. Instead, postphenomenological claims are posed from an embodied and situated perspective, refer to practical problems, and are empirically oriented. To both phenomenology and pragmatism, postphenomenology adds a focus upon case studies of human-technology relations. Through this combination of traditions and interests, the postphenomenological perspective has developed a unique set of insights into technology, and applies these ideas to practical cases of usage, design, policy, and scientific research.

The central mode of investigation for postphenomenology is the application and analysis of the framework of concepts developed by Don Ihde, the founding figure of this perspective. Over a long and continuing career, Ihde has adapted insights from the phenomenological tradition for use in the concrete description of human relations to technology, and has developed his own account of humanity's contemporary technological situation. Those working in the school of postphenomenology refine and expand on Ihde's framework, and they apply these ideas to a wide range of philosophical issues, including technological agency, ethics, selfhood, anthropological methodology, politics, philosophy of design, and scientific practice. Just a few of the concrete topics addressed by case studies in this perspective include scientific and medical imaging, computer interface, virtual reality, traffic safety, robotics, educational technologies, sustainable design, wearable computing, and bodily implants.

* * *

This book consists of four sections. While a first part introduces the main ideas of postphenomenology, the second and third parts of this book present paradigmatic examples of postphenomenological essays by scholars working at the cutting edge of this perspective. The second section focuses on theoretical positions in the field, presenting a number of original postphenomenological analyses, while the third section presents a selection of postphenomenological case studies. This distinction between theoretical and empirical contributions to postphenomenology is somewhat false, though, since the chapters on theory include many concrete examples, and the empirical chapters include deep discussions and critiques of theory. The book concludes with a fourth section, in which "critical interlocutors" offer advice for this developing school of thought.

***Part 1*** The first section introduces the main ideas and philosophical positions of postphenomenology and is written by the co-editors of this volume, Robert Rosenberger and Peter-Paul Verbeek. This chapter, which aims to function as a "field map," includes a review of the main concepts of the postphenomenological framework, an analysis of its philosophical commitments, a description of postphenomenological methodology, and also a review of several of the key case studies of this perspective.

***Part 2*** Part 2 begins with Lenore Langsdorf's "Why Postphenomenology Needs a Metaphysics." Langsdorf argues that since postphenomenology attempts to articulate a world of human-technology relations, it would benefit from explicitly developing a metaphysics of process, one that "provides a theory of how interrelation produces novel entities." For this project, she advocates an incorporation of the metaphysical insights of Dewey, Whitehead, and ethnomethodology.

In his chapter, "What Robotic Re-embodiment Reveals About Virtual Re-embodiment: A Note on the Extension Thesis," Kirk Besmer explores what it means to embody remote devices, such as tele-surgery and remotely operated vehicles, and also immersive technologies, such as virtual reality and advanced videogames. He develops the argument that both cases should together be understood as distinct from classical cases of embodiment (such as a hammer or glasses) in which bodily experience is extended through the device.

Aud Sissel Hoel and Annamaria Carusi argue for the incorporation of insights from the later work of Maurice Merleau-Ponty into the postphenomenological perspective in their chapter, "Thinking Technology with Merleau-Ponty." Building on his notion of flesh, Hoel and Carusi develop the notion of the "measuring body" to articulate the complex the roles of technology within the dynamics of human perception.

In her chapter, "Movies and Bodies: Variations of the Embodied Self in Science-Fiction Techno-fantasies," Marie-Christine Nizzi shows that science fiction can provide a rich resource for thinking about technology and human bodies. She reasons that since blockbusters are seen by millions of people, they may speak widely held concerns, and she considers the implications of a variety of films, from *Avatar*, to *Bladerunner*, to *Watchmen*.

Fernando Secomandi's chapter ("Bodies as Technology: How Can Postphenomenologists Deal with the Matter of Human Technique?") addresses the issue of human bodily training. In cases in which our trained bodily techniques are relevant to our relation to the world, Secomandi wonders if our bodies themselves should be considered technologies. He writes, "In the interpretation I propose, Ihde does not really convince us of any fundamental difference between technology and technique, but only reinforces their similarities."

Finally, in Asle Kiran's chapter, "Four Dimensions of Technological Mediation," he considers Ihde's claim that all technological mediation involves both a magnification and a reduction of some aspects of our experience. Building from this basic insight, Kiran outlines similar binaries at work in the ontological, epistemological, practical, and ethical dimensions of human-technology relations.

***Part 3*** In the first chapter of part 3, titled "Tracing the Tracker: A Postphenomenological Inquiry into Self-Tracking Technologies," Yoni Van Den Eede explores wearable computing devices that monitor the functioning of the user's own body. Through an analysis of the experience of these devices, Van Den Eede develops insights into the kind of self experienced through self-tracking. And he considers the implications of these devices—in which the target of new information is the user her or himself—for postphenomenology's conception of the mediation of technology.

The chapter "A Century on Speed: Reflections on Movement and Mobility in the Twentieth Century," by Søren Riis, investigates how the notion of speed is closely linked to technological developments. Riis threads together a variety of contexts and conceptions of speed, exploring everything from the history of early flight, to the Olympics, to amphetamines, to the accelerating rate of technological change.

In her chapter, "Searching for Alterity: What Can We Learn From Interviewing Humanoid Robots?" Frances Bottenberg investigates the experience of holding conversations with highly realistic robots, especially Bina48, a robotic bust designed for advanced verbal conversation. She reflects on the implications of these experiences for the future of human-robot conversational etiquette, and what these experiences ultimately imply about ourselves. Bottenberg writes, "Perhaps the culprit is vanity, perhaps lack of imagination, or perhaps simply our sociable predisposition, but we remain inclined to read ourselves—even our own personalities and faces—into the manufacturing of the significant AI others we face."

Chris Kaposy, in his chapter, "Postphenomenology of the Medical Student," draws parallels between contemporary medical ethics training and the programming of robots. He argues that the dominant approach, with its automation-like emphasis on systems and standardized rules, would benefit from a postphenomenological conception of multistable embodied subjects.

Adam Rosenfeld's chapter, "Mediating Multiplicity: Brain-Dead Bodies and Organ Transplant Protocols," cross-analyzes various conceptions of corpses ready for organ donation. Rosenfeld argues that, while there is a tendency to understand brain-dead bodies as simple Cartesian cadavers, it is beneficial to conceive of their multistability, and consider their different roles shifting through different practical organ donation contexts.

Jan Kyrre Berg Friis explores the concrete practices of image interpretation in the field of Radiology in his chapter, "Towards a Hermeneutics of Unveiling." Through an analysis of the details of these practices, Friis considers what it means for those working in postphenomenology to become immersed within the meaningful context of a practical area of empirical study.

As a supplement to these two middle sections of the book, we would like to direct readers to the forthcoming sister volume to this one, *Technoscience and Postphenomenology: The Manhattan Papers*, edited by Jan Kyrre Berg Friis and Robert P. Crease. This book contains many chapters that also represent the cutting edge of postphenomenological research by mainstay contributors to the panels and publications hosted by this school of thought, including Cathrine Hasse, Kyle Powys Whyte, Shannon Vallor, Anette Forss, Junichi Murata, Bob Scharff, Stacey Irwin, Galit Wellner, Gert Goeminne, and many others, including longtime interlocutors with Ihde. *The Manhattan Papers* and the current book were purposely organized together to showcase the work of many of the core contributors this perspective.

***Part 4*** The fourth section of this book features critical responses to the postphenomenological perspective by three interlocutors with this school of thought: Andrew Feenberg, Diane Michelfelder, and Albert Borgmann. The works of these three figures are of constant relevance and concern to those working on postphenomenology. Moreover, all three have recently engaged with postphenomenologists through journal issues and conference panels. In this section, they offer critical comments and advice for the school of postphenomenology, prescribing the future directions they would like to see this perspective take.

In his chapter, "Making the Gestalt Switch," Andrew Feenberg argues that much of his own work on the political and socially constructed dimensions of technology can be understood as consistent with the aims and commitments of postphenomenology. Thus, if postphenomenologists wish to expand their analyses into issues regarding the construction of technology through political, economic, and bureaucratic structures, then they can look to the insights of his instrumentalization theory. Feenberg explores how this can be done in part through an analysis of the work of Lukács.

Diane Michelfelder's chapter, "Postphenomenology with an Eye to the Future," argues that the postphenomenological perspective is well suited to the task of analyzing new technologies approaching on the horizon, such as wearable computing devices like Google Glass. But Michelfelder also develops the criticism that, in its concern with concrete empirical case studies, postphenomenological work fails to account for the experience of the world as a whole. If not expanded to both descriptively and normatively provide such an account, she argues that postphenomenology will be ill-equipped to address future technologies that will inform our relationships with the world itself, from biosensors to "smart" environments.

In his chapter, "Stability, Instability, and Phenomenology," Albert Borgmann reflects on the postphenomenological notions of multistability and variational theory, and considers examples from the fiction of Kees Boeke which include scenes that expand into outer space and shrink view down to the microscopic level. Borgmann reflects on what it means to take up the multiple perspectives adopted by postphenomenology, what must remain invariant in such analyses, and what it means for all this to take place in the context of moral horizons that relate to the cosmos, the micro-world, and the related global concerns. He writes, "How these horizons morally inform the lifeworld is one of the great unanswered questions of contemporary philosophy and requires a new phenomenology."

* * *

We sincerely hope that this book will contribute to the further development of the postphenomenological approach. By providing an overview of the basic characteristics of the approach, and by bringing together a substantial number of theoretically and empirically oriented postphenomenological analyses and critical perspectives, we hope to lay a basis for further investigation and discussion. Parallel to this book, the forthcoming volume *Technoscience and Postphenomenology: The Manhattan Papers*, edited by Jan Kyrre Berg Friis and Robert P. Crease, will present more state-of-the-art work in postphenomenological approach. This is exactly what we aim to accomplish with this book series in *Postphenomenology and the Philosophy of Technology*: to be a platform for high-quality, state-of-the-art work in postphenomenology, where Philosophy of Technology meets Science and Technology Studies, and where philosophical analysis meets the concrete materiality of science and technology, and—with a wink to Husserl—finds a new way "to the things themselves!"

# 1

# A Postphenomenological Field Guide

## *Chapter One*

# A Field Guide to Postphenomenology

Robert Rosenberger and Peter-Paul Verbeek

Over the past decades, an expanding group of scholars has been developing a novel approach to the social and cultural roles of technology. Building upon Don Ihde's phenomenological analyses of human-technology relations (e.g., Ihde 1990), they have started to study technologies with a blend of empirical and philosophical research methods—ranging from the epistemic role of the Mars explorer vehicle to the role of technologies in education and from the impact of hands-free calling on driving behavior to the role of sonography in moral decisions about abortion.

All of these studies label themselves as "post-phenomenological," in order to express their ambivalent relation to the phenomenological tradition. On the one hand, they are heavily inspired by the phenomenological emphasis on experience and concreteness, while on the other hand they distance themselves from the classical-phenomenological romanticism regarding technology, and find a starting point in empirical analyses of actual technologies.

The various "postphenomenological studies" that have been undertaken so far (e.g., Ihde, 1993; Selinger, 2006; Rosenberger, 2008; Verbeek, 2008b; Rosenberger, 2012) have at least two things in common. First of all, they all investigate technology in terms of the *relations between human beings and technological artifacts*, focusing on the various ways in which technologies help to shape relations between human beings and the world. They do not approach technologies as merely functional and instrumental objects, but as mediators of human experiences and practices. Second, they all *combine philosophical analysis with empirical investigation.* Rather than "applying" philosophical theories to technologies, the postphenomenological approach takes actual technologies and technological developments as a starting point for philosophical analysis. Its philosophy of technology is in a sense a philosophy "from" technology.

This blend of empirical and philosophical work has found itself a place in the existing repertoire of methods in the Philosophy of Technology and in Science and Technology Studies, bringing together the empirical orientation of STS on concrete case-studies with the conceptual and also normative orientation that are characteristic for philosophy of technology.

This chapter provides a field guide to postphenomenology. To be sure: there is no strict postphenomenological methodology that scholars could follow. Postphenomenology comes in just as many flavors as there are scholars in the field. Still, these postphenomenological flavors have some common features and characteristics. In order to discuss these, we will first explain how postphenomenology developed out of "traditional" phenomenology. After that, we will elaborate some of the central concepts in the approach, and discuss some elements of its methodology. We will conclude the chapter by discussing a number of postphenomenological case studies, which show concretely what the approach has to contribute to the study of science and technology.

## FROM PHENOMENOLOGY TO POSTPHENOMENOLOGY

The postphenomenological approach in philosophy of technology has developed out of a critical dialogue with the phenomenological tradition on the one hand and research in the empirical field of Science and Technology Studies on the other. Classical phenomenological analyses of technology, most notably in the work of Martin Heidegger, approached technology in fairly abstract and also romantic terms. They studied "Technology" as a broad, social, and cultural phenomenon, with a special focus on the ways in which technology alienates human beings from themselves and from the world they live in. While this approach has brought many relevant insights in the role of technology in human existence, its monolithic and romantic character were increasingly experienced as problematic (Feenberg, 2000; Ihde, 1993). The analyses were losing touch with the actual experiences people have of the roles of technologies in human existence.

At the same time, the empirical approach of the field of Science and Technology Studies does not always provide a real answer to this lack of connection. Despite its empirical basis in fields like sociology and anthropology, and despite its ambition to take an "empirical detour" to answer philosophical questions, it eventually did not always find the way back to these questions. Insights in the dynamics of the complex interactions of science, technology, and society, however valuable they are, do not always help to answer the philosophical question of how the role of technology in human existence and experience can be understood.

The postphenomenological approach combines an empirical orientation with philosophical analysis. It calls itself "post"-phenomenology to emphasize that it distances itself from the romanticism of classical phenomenology. Rather than positioning itself in opposition to science and technology, it aims to integrate science and technology in its analysis of the relations between human beings and their world. Classical positions have often claimed that phenomenology provides a rich alternative to the narrow scientific and technological approach to reality. While the sciences, as Merleau-Ponty stated, merely "analyze" things from a distance, phenomenology "describes" them from a closer engagement (Merleau-Ponty, 1962). Rather than describing the world as it actually is, the sciences present a reduced reality. Against this, phenomenology takes us "back to the things themselves."

It is precisely this claim to regain access to an original world that is richer in meaning than the world of science and technology, that postphenomenology refutes. Rather than thinking in terms of alienation, it thinks in terms of *mediation*. Science and technology help to shape our relations to the world, rather than merely distancing us from it. This perspective of mediation embodies a reinterpretation of the foundations of phenomenology. It does not see phenomenology as a method to *describe* the world, but as understanding the *relations* between human beings and their world. In fact, this study of human-world relations is what all major phenomenological thinkers have been doing, by conceptualizing it in terms of consciousness (Husserl), perception (Merleau-Ponty), being-in-the-world (Heidegger), and so on (cf. Verbeek, 2005, 99–119). Building upon and expanding these conceptualizations of human-world relations makes it possible to understand how science and technology mediate these relations, rather than merely impoverishing or suffocating them.

The "post" in "postphenomenology," however, does not only indicate that it moved beyond classical phenomenology. It is also reveals a remote relation to postmodernism. Without considering itself postmodern, postphenomenology expands the ways in which phenomenology has criticized modernism. In fact, phenomenology itself was a result of discontent with the modernistic separation of subject and object. Rather than seeking the source of knowledge in either subjective ideas or objective facts, it focused on the intentional relation between subject and object. Subject and object cannot have a separate existence. The human subject is always directed at objects: we cannot just "see," "hear," or "think," but we always see, hear, or think *something*. Similarly, the objects "in themselves" will probably exist, but it does not make much sense to think about them, because as soon as we do that, they become things-for-us, things as disclosed in our relations with them. Phenomenology replaced the split between subject and object with an *intentional relation* between them.

Postphenomenology takes this relationist approach one step further than phenomenology. Phenomenology itself was already a move beyond modernism, because its concept of intentionality made it possible to overcome the modernistic subject-object split. Against subjectivism and objectivism, which founded knowledge on the internal workings of the subject or on the objectivity of the world, its relationalism opened a novel and fruitful perspective. Postphenomenology, however, reconceptualizes this intentional relation in two distinct ways. First, it investigates its fundamentally mediated character. There is no direct relation between subject and object, but only an "indirect" one, and technologies often function as mediators. The human-world relation typically is a human-technology-world relation (Ihde, 1990). Second, it does away with the idea that there is a pre-given subject in a pre-given world of objects, with a mediating entity between them. Rather, the mediation is the *source* of the specific shape that human subjectivity and the objectivity of the world can take in this specific situation. Subject and object are *constituted* in their mediated relation (Verbeek, 2005). Intentionality is not a bridge between subject and object but a fountain from with the two of them emerge.

This focus on mediation and mutual constitution sharply demarcates the postphenomenological approach from classical phenomenology. Claiming a privileged access to the things themselves becomes an impossibility within the postphenomenological approach. And against the idea that technology alienates human beings from the world and from themselves it places the idea that technologies help to shape human subjectivity and the objectivity of the world. Postphenomenology makes it possible to make micro-scale analyses of the mediating roles of technologies in human-world relations—and as such it can be said that it truly takes us back "to the things themselves": material technological artifacts that deserve explicit philosophical attention.

This turn towards materiality and concrete practices reveals an influence of American pragmatism. It is in practices of interacting with technologies where the phenomenon of technological mediation occurs and can be studied. Human-world relations are practically "enacted" via technologies. Galileo's moon, to use an example of Don Ihde, took shape in the complex, practical interactions of Galileo with his telescope, on the basis of which he came to develop a new interpretation of what he saw through his device. The materiality of technology can be studied best in concrete, practical situations of use. This is beautifully reflected in the Greek word "pragmata," which means "things," but which is closely related to the word "praxis," which means "practice." Postphenomenology is the practical study of the relations between humans and technologies, from which human subjectivities emerge, as well as meaningful worlds.

As a result of this practical and material orientation, postphenomenology always takes the study of human-technology relations as its starting point. In order to understand a technology or a technological development, postphenomenology always analyzes the character of the relation human beings have with this technology and the ways in which it organizes relations between human beings and the world. Human beings can interact with technologies, incorporate them, read them. All of these relations organize how human beings experience their environment, and how they are practically engaged with it. Technologies, to be short, are not opposed to human existence; they are its very medium.

## KEY CONCEPTS

Work in the postphenomenological perspective most often proceeds through the application of its framework of original concepts to specific cases of human-technology relations. Many of these ideas were first developed by Ihde, and have been refined throughout his career. And much of this framework has been cultivated and expanded by others. In this section we review some of the most influential ideas of the postphenomenological framework of concepts.

### Human-Technology Relations

One of Ihde's most influential contributions to the philosophy of technology is his breakdown of the different forms that human-technology relations may take. This list of the various ways that users can establish bodily-perceptual relationships with technologies has proven useful in several ways. First, these concepts have practical value for case studies of user relations to technologies. That is, this categorization of different forms of technological mediation has proven useful in the concrete description of user relations to technologies. Second, the list itself demonstrates the variability and context specificity of user experiences. As such, this list argues against oversimplifying or one-size-fits-all accounts of user experience. And lastly, the conception of human-technology relations reflected in this list is beginning to buckle at its edges in productive and provocative ways. The issues arising at the boundaries of this overall conception of the forms of technological mediation represent some of the most exciting philosophical frontiers of postphenomenology in particular, and the field of philosophy of technology more generally.

In what follows in this subsection, we summarize Ihde's four basic forms of technological mediation: embodiment relations, hermeneutic relations, alterity relations, and background relations.

*Embodiment Relations*

While this list of the forms of human-technology relations should not be understood as exhaustive, taken together they help to articulate many of the ways that users develop bodily-perceptual relationships with the devices they use. With the notion of "embodiment relations," Ihde points to the mediation of those technologies which transform a user's actional and perceptual engagement with the world. When a technology is "embodied," a user's experience is reshaped *through* the device, with the device itself in some ways taken into the user's bodily awareness. (While Ihde develops these ideas throughout his corpus, the versions reviewed here come primarily from chapter 5 of his seminal *Technology and the Lifeworld* [1990].)

Ihde provides the following diagrams. If the technological mediation between a user and the world can be represented as follows,

Human—Technology—World

then an embodiment relation to technology can be represented in this manner:

(I—Technology) → World

The go-to example is a pair of eyeglasses. As they are worn, the glasses themselves are not simply one among the many things in the world the user may perceive. Instead, a user looks *through* the glasses upon a transformed world, and the glasses can be conceived as a part of the user's perceptual experience. Ihde writes, "the wearer of eyeglasses embodies eyeglass technology: I—eyeglasses—world" (1990, 73). As they are worn, the eyeglasses are a transformative mediation of the bodily-perceptual relationship between the user and the world.

A related notion is what Ihde calls the "transparency" of a particular human-technology relation. This refers to the degree to which a device (or an aspect of that device) fades into the background of a user's awareness as it is used. As a user grows accustomed to the embodiment of a device, as habits of bodily action and perception develop, and as the use of the device takes on a familiar and everyday character, the device itself takes on a degree of transparency. Ihde writes, "My glasses become part of the way I ordinarily experience my surroundings; they 'withdraw' and are barely noticed, if at all. I have then actively embodied the technics of vision" (1990, 73). Ihde goes so far as to say that with regard to the design and use of technology we maintain a "double desire": we want a technology to at once both optimally transform our relationship to the world, and at the same time we want the experience of the means of that transformation to itself remain as experientially transparent as possible (1990, 75). (For a recent discussion on the notion of transparency, see Van Den Eede, 2011.)

The notions of embodiment and transparency can be illustrated in the contrast between two wearable technologies with which Ihde himself is personally familiar, hearing aids and eyeglasses. The two can be contrasted in terms of the kind and degree of transparency possible to achieve with each:

> My awareness of wearing glasses is a fringe awareness that gets interrupted only when there is back glare, or when the glasses slip off my nose, or when the lenses get dirty or smudged, when, in other words, something diminishes the normative transparency of the optics. . . . With hearing aids, however, the technology of interest is an acoustic or auditory technology, a hearing "aid" ideally should function parallel to the visual eyeglasses example. Unfortunately the auditory transparency is much more difficult to attain, a fact well recognized by audiologists and others. A significant number of people attempt to use hearing aids, but the difficulty of embodiment is sometimes such that they give up. (Ihde, 2007, 244–245)

According to Ihde, a number of specific features of hearing aid technology account for this contrast in the level of transparency possible for glasses and hearing aids, from the stuffiness of the feeling of the device wedged into the ear, to the devices' inability to distinguish between local and background sounds for amplification.

It would be not be accidental if the reader were to detect in the notions of embodiment and transparency an echo of Heidegger's account of tool use (as well as, of course, other examples from the phenomenological canon that share these themes, such as Merleau-Ponty's account of the blind man's cane). Ihde's framework on this point is at once an appropriation and critique of Heidegger's ideas. That is, on the one hand Ihde's conceptions of the embodiment and transparency of technology can be understood as a straightforward repackaging of Heidegger's description of the withdraw of the ready-to-hand tool, such as in his famous account of the hammer in *Being and Time* (1996). Where Heidegger's account is a critique of the entire history of Western metaphysics, within Ihde's repackaging these ideas are instead put toward the pragmatic description of particular, context-dependent, human-technology relations. On the other hand, Ihde's account of embodiment relations is part of an explicit critique of Heidegger's views. Crucially, the notion of embodiment relations in Ihde's framework is only one of several potential forms of mediation. Additionally, the particular degree to which a device should be considered transparent within an embodiment relation will depend on the case. Thus, against Heidegger's (allegedly totalizing) account of the metaphysics of tool use, Ihde's notions of embodiment and transparency are part of a larger account designed specifically to emphasize the variability and context-dependency of human-technology relations (see, e.g., 1990, 80, 98).

According to Ihde, human-technology relations generally, and embodiment relations in particular, share a magnification/reduction structure. He writes, "Embodiment relations display an essential magnification/reduction structure which has been suggested in the instrumentation examples. Embodiment relations simultaneously magnify or amplify and reduce or place aside what is experienced through them" (Ihde, 1990, 76). That is, through the non-neutral transformations rendered to user experience through the mediation of a technology, we not only receive the desired change in our abilities, but always also receive other changes, some of them taking on the quality of "tradeoffs," a decrease of a sense, or area of focus, or layer of context. A hammer enables one to strike a nail with force, but occupies one's hand in the process. A pair of binoculars enables one to see over a distance, but at the sacrifice of visual awareness of one's immediate surroundings. A telescope image of the moon, Ihde notes, enables us to see its surface in great detail, but removes the moon from its context in the sky (1990, 76).

Ihde also argues for a distinction between what he calls microperception and macroperception (1993, 74; see also the notions of "Body 1" and "Body 2" in Ihde, 2003). Where "microperception" refers to the individual bodily sensations articulated in the work of Husserl and Merleau-Ponty, "macroperception" instead refers to the cultural, historical, and anthropological dimensions of experience explored by figures such as Heidegger and Foucault. While it can be helpful at times to analyze these two dimensions of experience separately, Ihde holds that they remain inextricable. He writes, "The matter may be put simply: there is no bare or isolated microperception except in its field of hermeneutic or macroperceptual surrounding; nor may macroperception have any focus without its fulfillment in microperceptual (bodily-sensory) experience" (Ihde, 1993, 77). Defenders of these concepts argue that, where historically phenomenology has often failed to address cultural dimensions in the study of experience, these notions highlight these issues and urge for their indivisibility with studies of the structure of experience. But critics contend that cleaving these dimensions into distinct concepts threatens to relegate issues of culture, gender, race, and difference into a separate conversation (e.g, Eason, 2003; Scharff, 2006). The practical value of the micro/marcoperceptual distinction can be seen in recent postphenomenological case studies, for example finding a central role in anthropological work on laboratory practice (e.g., Hasse, 2008; Forss, 2012).

### *Hermeneutic Relations*

The second major form of technological mediation that Ihde identifies is called "hermeneutic relations." Pointing to the hermeneutic tradition of philosophy (which contemplates the nature of language interpretation and translation), Ihde uses the notion of "hermeneutic relations" to refer to technologies which are used through an act of perceiving and interpreting the device's readout. Rather than experience the world through the device as in an embodiment relation, in a hermeneutic relation the user experiences a transformed encounter with the world via the direct experience and interpretation of the technology itself. Ihde provides the following representation for hermeneutic relations:

$$I \rightarrow (\text{Technology—World})$$

A wristwatch is a simple example. A user looks at the watch's face, interprets the hands or digital display, and through this hermeneutic relation experiences a transformed access to the precise time of day.

In the case of hermeneutic relations to technology, the analogy is between reading written language and "reading" the display interface of the mediating technology. A key aspect of one's encounter with a written language is the degree to which one "knows" it, that is, the degree to which one understands how to read that particular language. If a person has no experience with that language, words written in it will convey no meaning at all. But if she or he instead is deeply fluent, then the meanings of words jump from the page almost automatically. For the person deeply familiar with a language, she or he does not need to slowly and actively interpret each letter, then put them together to make a word, and then interpret the meaning of that word; the meaning emerges all at once in a perceptual gestalt. This is also true for hermeneutic relations to technology. In the case of the wristwatch, transformed access to the time of day will only appear for a wearer that already "knows how to tell time." But for a wearer familiar with the reading of clocks, the precise time of day emerges from the wristwatch all at once in a perceptual gestalt.

The transparency of a particular hermeneutic relation will depend on the particular level of familiarity one brings to the interpretation of the device's readout. Take, for example, a complex medical imaging technology such as an fMRI. For a new medical student entirely novice with regard to this technology, she or he may need to concentrate and slowly decode the meaning of the colorful brain-shaped display. But for the expert, much of the content and context of the readout will appear in a perceptual gestalt, and she or he can focus attention on the area of interest within the image.

It should be noted that, while many standard examples of hermeneutic relations focus on technologies with visual readouts, this is not a necessary feature of this notion. The device readout could potentially occur in terms of any bodily sense. For example, any number of devices can provide audio readouts which require direct perception and active interpretation, including the clicks of Geiger counters, the steady boops of heart monitors, the ca-ching of cash registers, and the squeal of a car's aging brake pads.

*Alterity Relations*

With the notion of "alterity relations," Ihde refers to devices to which we relate in a manner somewhat similar to how we interact with other human beings. The idea is that some forms of interface are devised specifically to mimic the shape of person-to-person interaction, and that sometimes we encounter a device as itself a presence with which we must interrelate. The term "alterity" is used in phenomenological discussions to refer to the special experience of engaging with another human being, that significant encounter with Otherness. Ihde claims that our interactions with technologies can sometimes take on a "quasi-other" quality. Ihde formulates the diagram for alterity relations as follows:

$$\text{I} \rightarrow \text{Technology} - (- \text{World})$$

One common form is computer interface schemes that pose direct questions to the user, such as the ATM machine that displays questions on its screen ("Would you like to make a withdrawal?"), or the "dialog box" that opens on a computer screen to provide program installation instructions. This is not to claim that we mistake these devices for actual people, but simply that the interface modes take an analogous form.

It could be predicted that as computing advances, and as our abilities to create sophisticated computer programs that simulate human interactive style increase, we will see more and more devices designed with an alterity-style interface. It is already the case that this form of relation can be increasingly seen in automated interactive customer service phone calls, GPS devices that read aloud driving directions, and voice interactive personal assistant smartphone applications. Accordingly, postphenomenologists continue to develop this idea (e.g., Irwin, 2006; Bottenberg, 2015; Wellner, 2014b).

*Background Relations*

With the notion of "background relations," Ihde addresses those technologies that make up the user's environmental context. A user shares a background relation to devices that she or he may not directly use but nonetheless interacts with as they shape her or his experiential surroundings. Examples include the running refrigerator on the other side of the room, or the central air conditioning and/or heating system that goes on and off automatically throughout the day.

While it may be tempting at first to understand the lack of attention paid to background relations in terms of technological transparency, Ihde clarifies this point. He explains, "The 'withdraw' of this technological function is phenomenologically distinct as a kind of 'absence.' The technology is, as it were, 'off to the side.' Yet as a present absence, it nevertheless becomes part of the experienced field of the inhabitant, a piece of the immediate environment" (Ihde, 1990, 109). Technologies to which we share background relations stand back in our awareness not simply because we have grown accustomed to their usage, but because they quite literally form the backdrop of our experiences. They shape our experiences, protecting us from the elements or keeping our food safely chilled, but do so in ways that do not require direct interaction.

## Relational Ontology

The postphenomenological focus on human-technology relations, as stated in section 1, is closely connected to its relational ontology. Technologies, postphenomenology holds, are to be understood in terms of the relations human beings have with them, not as entities "in themselves." When technologies are used, they help to establish relations between users and their environment: a telescope organizes a relation between an observer and a heavenly body, just as an ultrasound device helps to shape a relation between expecting parents and their unborn child. In doing so, technologies also help to shape the "subjectivity" of their users and the "objectivity" of their world: telescopes constitute their users as "observers" and the sky as "observable," just as ultrasound constitutes the unborn child as a potential patient and expecting parents as those who are responsible for the health condition of their child. Subject and object are no pre-given entities, but get constituted in the technologically mediated relations that exist between them.

This relational ontology of the postphenomenological approach has a somewhat different character than the ontology of the closely related approach of Actor-Network Theory (Latour, 1993). Actor-Network Theory approaches the world as networks of relations between "actants," which can be human and nonhuman. Latour emphasizes that his approach of these actants is "symmetrical": he does not want to start from an *a priori* distinction between human subjects and nonhuman objects, but rather aims to make visible the continuity between humans and nonhumans. Revealing this continuity, according to Latour, makes it possible to understand how nonhuman entities do not only play a role in the material world, but in the social world as well. From such a symmetrical approach, not only humans but also things "act."

The postphenomenological approach, however, explicitly does not give up the distinction between human and nonhuman entities. Instead of symmetry it sees interaction and mutual constitution between subjects and objects. Just like in Latour's approach, there are no pre-given subjects and objects;

subjectivity and objectivity are always the *product* of relations, rather than their starting point. Things are not symmetrical to humans, but together, humans and things constitute all kinds of entities. In order to see these processes of mutual constitution, and to do justice to human experiences of being subjectively "in" a world, it remains very relevant to make a distinction between humans and things. When we give up this distinction, we also give up the phenomenological possibility to articulate (technologically mediated) experiences "from within." Actor-Network Theory studies complicated networks of relations "from outside," from a third-person perspective; postphenomenology studies engaged human-world relations, and their technologically mediated character, from a first-person perspective. It is not the *distinction* between humans and technologies that it wants to depart from, but their radical *separation* (see Verbeek, 2005, 166–168).

From this subtle difference between "separating" humans and nonhumans on the one hand and "distinguishing" them on the other, it also becomes possible to conceptualize the "active" role of technologies. When postphenomenology claims that technologies play an actively mediating role in human-world relations, it does not claim that things can act just like humans do. Such a claim would actually reproduce the modernistic subject-object split, by attributing the characteristic of subjects to objects as well. The question is not: is agency not only a property of subjects but also of objects? Rather, the question is: what kind of roles do objects play in agency? Agency, then is not an exclusively human property anymore: it takes shape in complicated interactions between human and nonhuman entities.

## Cyborg Relations

Recent work in postphenomenology has expanded the range of human-technology relations that have been analyzed in Ihde's work. Verbeek has argued that postphenomenology should also be able to account for technologies like brain implants, domotics, and augmented reality devices, relations in which technologies start to merge either with the human body or with the environment (Verbeek, 2008a). The spectrum of four human-technology relations identified by Ihde is primarily relevant for analyzing human-technology configurations in which technologies are *used*. But at the extremes of this spectrum, other configurations become visible.

In the spectrum from embodiment via hermeneutic and alterity to background relations, technologies move ever "further away" from the human being as it were: from being an extension of the body via a "readable" artifact and an object for interaction to the background of our experience. More intimate than a relation of embodiment, though, there is a relation of "fusion," in which the physical boundaries between humans and technologies are blurred, and technologies merge with our bodies. And more radically "environmental" or "ambient" than the background relation, there is a relation of "immersion," in which a technological background interacts actively with human beings. Moreover, augmented reality technologies, which add an extra layer to our experience of the world, open up a relation of "augmentation."

In all of these new human-technology relations, new "intentional relations" between humans and the world come into being. The concept of intentionality plays a central role in the phenomenological tradition, as part of its relational approach, to conceptualize the relation between human beings and their world. Human experience has an intentional structure: human beings are always directed toward reality. We cannot simply "see," but we always see something; we cannot simply "hear" but always hear something; et cetera. We cannot understand human experience without taking into account this intentional directedness towards the world. In all of the human-technology-world relations Don Ihde analyzes, technologies mediate this intentionality. In the new human-technology relations identified above, the technologically mediated character of intentionality takes on a different shape.

In the "fusion" relation, technologies merge with our physical body. Good examples are neuro-implants for deep brain stimulation; cochlear implants that enable deaf people to hear again; or artificial heart valves and pacemakers. In all of these cases, the relations between humans and technologies are more intimate than in the embodiment relation. They involve much more than wearing a pair of glasses, or listening through a cell phone. Lucie Dalibert even speaks of "somatechnologies": technologies that blur the boundaries between body and artifact (Dalibert, 2014). This cyborg relation can be schematized as:

(I / Technology) → World

From this "fusion" relation, a "hybrid intentionality" emerges. Rather than being a technologically mediated form of human intentionality, it is the intentionality of a new, hybrid entity: a cyborg. Contrary to embodiment relations, no clear distinction can be made here between the human and the nonhuman elements in these relations: together, humans and technologies form a new entity, which is directed towards the world in a "hybrid" way, integrating human and nonhuman elements.

In the "immersion" relation, the configuration of humans and technology takes yet another shape. Here, technologies do not merge with the body but with the environment. And the relation between human beings and that hybrid environment is interactive: smart environments "perceive" their users too, and "act" upon them. Examples are "smart toilets" that automatically analyze one's excrements and produce a medical report on the basis of this, or "smart beds" in hospitals that can detect if somebody falls or steps out of bed. Schematically:

I ↔ Technology/World

The intentionality that comes with this type of relations has a "bi-directional" character: human beings are directed towards technologies that are also directed towards them. This bi-directionality can have several implications. On the one hand, it makes it possible for human beings to experience how technologies "experience" them, resulting in a "reflexive intentionality": smart toilets open a new relation to oneself. On the other hand, it opens the possibility to be immersed in how smart environments perceive and act upon human beings, as is the case for some "persuasive technologies" that give unsolicited feedback on one's behavior. An example is the "persuasive mirror" that was designed for waiting rooms of medical doctors to give people feedback on their lifestyle. When looking in the mirror, you actually look into a camera and see yourself on a screen, but a face recognition system that is linked to your medical record makes it possible for the system to morph your face into what you will look like in five years' time if you don't give up smoking, eating too much, or working too hard. People who look into this mirror are immersed in the mirror's intentionality towards specific aspects of themselves, resulting in an "induced" intentionality.

The relation of augmentation, to conclude, adds a second layer to our world, a second field of attention to be intentionally directed to. In addition to the sensory relation we have with the world "through" such an augmented reality technology, we also have a relation to the information it gives us. Google Glass is a good example of such an augmented reality technology. People wearing it can get information on objects and buildings they see, and potentially also about human beings, when the technology would be equipped with face recognition technology; they can exchange messages, take pictures and record videos, and surf the internet "in the background" of their activities. When using Google Glass, people both have an embodiment relation with the Glass itself, and a hermeneutic relation with its screen that offers a representation of the world. Therefore it offers not one, but two parallel relations with the world. Schematically:

( I - Technology) → World
↘ (Technology – World)

The intentionality involved in such "augmentation relations" can be indicated as "bifurcated": there is a split in people's directedness at the world, because two parallel fields of attention emerge.

## The Field of Awareness

Rosenberger has identified the notion of technological transparency as a key aspect of the postphenomenological framework ripe for re-examination and expansion. Recall that "transparency" refers to the degree to which a device recedes into the background of a user's awareness as it is used. At times in Ihde's corpus, the notion of transparency is even cast as a defining feature of embodiment relations. But Rosenberger argues that this idea should be understood as only one of many potential ways that technological mediation shapes the contours of a user's overall experience. Postphenomenologists could investigate what stands forward in addition to what withdraws, what demands attention, what remains on the fringes, and just how strongly these particular features hold within a given relation. Rosenberger's contention is that transparency should be understood as only one feature among many that could characterize a user's experience within a given human-technology relation. This raises the question: how do different technologies reshape a user's overall "field of awareness" in different ways? Answering it will involve the development of a postphenomenological field theory.

Rosenberger develops two further variables that, like the notion of transparency, could characterize a user's technologically mediated field of awareness, what he calls "field composition," and "sedimentation" (e.g., Rosenberger, 2012; 2014b). This list need not (and should not) be considered comprehensive. But it helps point the way toward how a phenomenological field theory could be further developed.

### *Field Composition*

One alternate variable developed by Rosenberger is called "field composition." A human-technology relation deeply characterized by field composition is one in which a user's field of awareness is somehow significantly reconfigured by technological mediation. Rosenberger's inspiration here is the work of classical phenomenologist Aaron Gurwitsch who attempted to articulate a "field theory" for phenomenology (1964). Gurwitsch described the different ways that experience can be "organized," and developed an account of our overall "field of consciousness." (Rosenberger uses the term "field of awareness" to distinguish his own postphenomenological ideas from Gurwitsch's "field of consciousness.") While not subscribing to the specifics of Gurwitsch's account, Rosenberger argues that it is useful to attempt to articulate the ways that technological mediation reorganizes a user's field of awareness—reorganizations that could involve more than only the device's transparent withdraw. That is, it is useful to describe the specific ways a user's overall field of awareness is composed by technological mediation.

An example is a pair of binoculars. Not only is a user's bodily perceptual capacity to see across a distance greatly transformed, and not only may the binoculars themselves grow transparent in an accustomed user's hands, but the user's entire field of vision is altered. When looking through the binoculars, the user's field of awareness becomes filled by the content brought forward by the device, and the totality of user's vision is framed by a circle of darkness.

The experience of watching a movie in a theater is an even more encompassing example of a human-technology relation that is typically highly characterized by field composition. As the theater darkens, as the movie begins, and as the viewer becomes engrossed in the story, the viewer's awareness overall becomes almost entirely composed by the film's content. It seems insufficient to say that the surrounding theater, and the seat below, and the distance between the viewer and the screen all take on a degree of transparency. It is instead more appropriate to say that the movie content stands positively forward, its visual and audio content colonizing the user's field of awareness. Put differently, through the technological mediation of the movie theater, the viewer and the world are co-shaped such that (at least in the most engrossing moments) the movie content itself composes the entirety of the world as experienced. That is, the movie-watching experience is a human-technology relation highly characterized by field composition.

A rich example can found in the ethnographic work of Anette Forss on the laboratory practices of cervical cancer screening in Sweden (2005; 2012). Forss conducts postphenomenological studies of the procedures involved in analyzing pap smear samples under the microscope. As she observes, "Basically, the microscope was the tool that enabled them to "enter another world" and seemed for many of them to have become an extension of their bodies" (Forss, 2005, 72). The microscopists' usage of this device can be conceived as an embodiment relation to technology, and it would be possible to consider the transparency that develops with regard to the microscope interface, the withdraw of the interface features such as the eyepiece or focusing knobs. It is also possible to explore the hermeneutics of the interpretation task. For example, Forss describes the experience of scanning through samples. If abnormal cells are discovered, this perception occurs within a striking gestalt; the anomalous cells jump into view—or "poppar up" in Swedish (Forss, 2005, 112). But a striking aspect of Forss's description of the microscopists' experience is the way they are described to "enter another world" through the device. In this way, the experience of the microscope can be characterized as having a high degree of field composition; the user's field of awareness is organized by the microscope, composed almost entirely by the "world" within the slides. A dramatic instance can be seen in the case of a microscopist describing the experience of encountering an anomalous cell. "I am so occupied with what I have found that, that everything around me disappears" (Forss, 2005, 123). The entirety of her field of awareness is composed by the content of the findings.

*Sedimentation*

The notion of sedimentation is used throughout the phenomenological tradition to point to those past experiences settled in one's mind which actively contextualize present experience. For example, Merleau-Ponty writes, "there is a 'world of thoughts,' or sediment left by our mental processes, which enables us to rely on our concepts and acquired judgments as we might on things there in front of us, presented globally, without there being any need for us to resynthesize them" (1962, 130). The metaphor is of course to rock formations built up over time through the accumulation of small deposits. But unlike rock formations, this world of thoughts is not static; it actively informs our ongoing experience. Sedimentation provides the pre-perceptive context that enables our current perceptions to occur with immediate meaningfulness.

Rosenberger uses the notion of sedimentation to refer to the force of habit associated with a given human-technology relation. That is, a relation that is highly sedimented is one that is steeped in long-developed bodily-perceptual habits. To return again to the example of an embodiment relation with eyeglasses, for the accustomed user the glasses are experienced with a high degree of transparency. But this transparent relation can also be said to be highly sedimented, with that transparent character taken on with an immediacy and stubbornness that reflects the strength of the user's entrenched bodily habits. Another example is the expert reader taking up a hermeneutic relation to an fMRI readout. As the expert gazes upon the image, due to the pre-perceptive context of meaning developed and sedimented through years of training and experience reading this kind of display, the image's meaning appears all together in a perceptual gestalt.

## Multistability

Across multiple fields of thought (including, but not limited to, STS, design, and especially the philosophy of technology), a central question is how to conceive of the non-neutrality of artifacts. That is, how should we understand the way that technology at once in part determines our choices and actions, and yet at the same time itself remains open to our manipulation and interpretation? How is technology both something we design and use for our own purposes, and also something that influences, restricts, leads, inclines, or controls us? One of postphenomenology's main responses to this question can be found in the notion of "multistability." This refers to the idea that any technology can be put to multiple purposes and can be meaningful in different ways to different users. As Ihde puts it, "No technology is 'one thing,' nor is it incapable of belonging to multiple contexts" (1999, 47). Crucially, the notion of multistability simultaneously points to the fact that the materiality of the device constrains the potential relations to only certain uses and meanings. That is, a technology cannot mean simply anything or be used to do simply anything; only some relations prove experientially stable. In this vocabulary, a multistable technology has multiple "stabilities" or "variations" (these two terms are used interchangeably in this literature).

Ihde first develops the notion of multistability in his early-career work on visual perception. In his book *Experimental Phenomenology*, he considers the experience of someone encountering and interpreting a variety of simple visual illusions, that is, drawings that can be coherently interpreted in more than one way (second printing 1986; expanded second edition 2012). These simple examples are helpful for illustrating the ways in which our perception is trained, and how our perceptual expectations enable our visual experience to occur in term of gestalts. The standard example is the Necker cube visual illusion (see figure 1.1). The common understanding of the drawing is that it can be interpreted in more than one way, either as a three-dimensional box with the topmost side on the upper right, or as another box with the topmost side on the lower left.

Ihde uses the illusion to illustrate the multistability of visual perception, explaining that each cube shape can be understood as a separate stability. When trained to recognize these separate stabilities, each appears individually in terms of a visual gestalt. Ihde continues his examination of the cube, explaining that, "If a background story can be found to allow [a] third variation to appear, its stability will coalesce almost immediately" (1986, 96). He explores several additional stabilities possible for the drawing, including a simple two-dimensional hexagon, a square-shaped bug straddling a hexagonal hole, and a cut gem shape in which the center square is taken as topmost with the surrounding sides sloping away. Ihde has also explored the multistablity of auditory phenomena (2007).

Despite being first developed as an account of human perception, the notion of multistability has come to be applied to human-technology relations. Artifacts are understood to potentially support multiple embodiment relations or hermeneutic relations (or other relations). A technology that supports multiple stable embodiment relations is one which could offer multiple potential transformations of a user's bodily-perceptual encounter with the world. Expanding on Heidegger's hammer analysis, Ihde claims,

> Heidegger's hammer is a simple example: a hammer is "designed" to do certain things—drive nails into the shoemaker's shoe, or into shingles on my shed, or to nail down a floor—but the design cannot prevent the hammer from (a) becoming an objet d'art, (b) a murder weapon, (c) a paperweight, etc. Heidegger's insight was to have seen that an instrument *is what it does, and this in a context of assignments*. But he did not elaborate upon the multistable uses *any* technology can fall into with associated shifts in the complexes of "assignments" as well. (1999, 46–47)

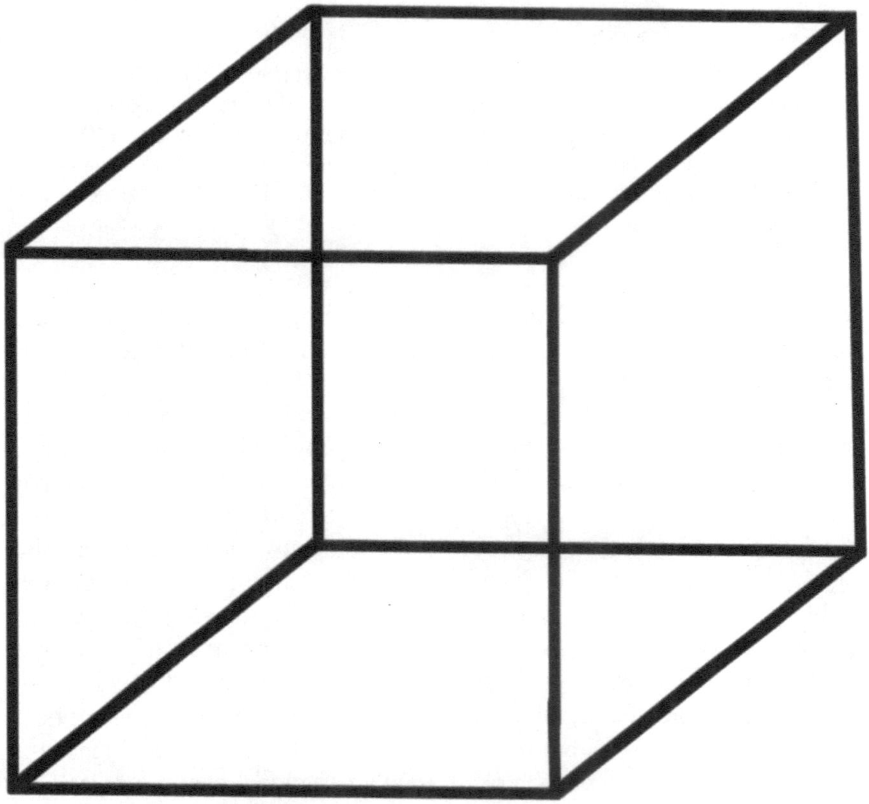

**Figure 1.1.**

A technology that supports multiple stable hermeneutic relations is one that a user can potentially perceive and interpret in different meaningful ways. An example would be two doctors disagreeing about the implications of a medical image (e.g., an X-ray, or a CAT scan) for a patient's diagnosis.

The term "variational analysis" refers to the method of brainstorming stabilities of a multistable technology. The idea of variational analysis comes straight out of the work of Edmund Husserl, but is crucially altered for use in the postphenomenological framework. For Husserl, variational analysis is a method in which an object of study is observed from multiple perspectives for the purpose of identifying features that remain present from any vantage point. According to Husserl, this process reveals the object's essence, its essential features free from the contingencies of perspective (1973). In contrast, postphenomenology—with its pragmatic commitments to anti-essentialism and context-dependent knowledge—rejects the idea that variational analysis enables one to discover the essence of an object of study. Ihde explains that,

> In Husserl's earlier use, variations (originally derived from mathematical variational theory) were needed to determine essential structures, or "essences." Variations could be used to determine what was variant and invariant. I have also found this technique invaluable in any phenomenological analysis—but as I used this technique, I discovered something other than Husserlian "essences" as results. What emerged or "showed itself" was the complicated structure of *multistability*. (2009, 12)

So for the postphenomenologist, the brainstorming of a technology's multiple stabilities serves to highlight technology's very context-dependent and materially-situated relationality.

The notion of multistability is put to an assortment of uses within postphenomenological studies. These can be generally divided into two categories: negative uses, and positive uses (Rosenberger, 2014a). In the "negative" use of multistability, a technology is subjected to variational analysis as part of a critique of another account seen to be somehow totalizing, essentializing, or overgeneralizing. The identification of a technology's status as multistable is taken to counter that totalizing account by showing the technology's variability. The quote above in which Ihde brainstorms various alternative possible stabilities for the hammer (the murder weapon, the objet d'art, etc.) is an example of a negative critique, a coy shot across Heidegger's bow; insofar as Heidegger holds that a hammer can only ever be experienced as transparently embodied or as broken, Ihde's variational analysis reveals the hammer's unappreciated multistability.

The notion of multistability can also be used as a part of "positive" research projects which seek to uncover new information about concrete cases of human-technology relations, or to reinterpret cases in productive ways. In such projects, variational analysis is taken up not simply to demonstrate that a technology is in fact multistable (as in the negative usage), but to investigate alternative stabilities to the "dominant" one, and to consider what these alternatives might teach us about the dominant stability. These projects also explore the ways that technologies get taken up through multiple stabilities in actual practice, and consider the implications of multistability for topics such as design, user training, and scientific research. An example is Rosenberger's series of papers on the multistability of user relations to computer interface (e.g., 2009; 2013b; 2014b). In one study, he considers the experience of a user sitting in front of a computer which has suddenly stopped working properly, such as when an Internet webpage loads with unexpected slowness. An exploration of alternative possible stable relations to the computer highlights the learned preconceptions and habits of practice that characterize typical human-computer interaction.

Another example is Cathrine Hasse's anthropological studies of health care practices regarding "Paro," a soft, robotic, seal-shaped device (2013). The device is an emotionally assistive and responsive "pet" designed in Japan to help the elderly, especially those with Alzheimer's. Hasse considers the multistability of this small robot with regard to how it has found different meanings within different cultural contexts; where in Japan the device more easily fulfills its emotionally supportive role, in Denmark the results are mixed. In the different cultural context of Denmark, sometimes the device is dismissed as a children's toy, it sometimes induces anger, and it only sometimes has soothing effects. Hasse also explains that, "Paro as a mediating artifact assumed its intended role as a tool for soothing Alzheimer's patients, but it also represented multistable signs for different people engaged in the activities of the nursing home" (2013, 95). Paro has proven disruptive as well for those working in the nursing home context, reorienting concrete practices, and challenging conceptions of care itself.

In an effort to spotlight the human side of the relation between a user and a multistable technology, Rosenberger has developed the notion of "relational strategies" (e.g., 2009; 2013b; 2014a). A relational strategy is the particular understanding and bodily approach that enables a user to relate to a technology in terms of a particular stability. We can ask: what sort of conceptual, comportmental, and bodily-habitual approach is necessary for a user to take up a technology in terms of one particular strategy rather than another? If we return to Ihde's variational analysis of the hammer, we can consider what relational strategy would correspond to the dominant stability, and what would instead be involved with the alternative stabilities that Ihde has identified. That is, we could consider how the hammer is taken up in terms of the dominant stability—i.e., normal, everyday nail hammering—and examine how it is explicitly and implicitly conceived, review its use context, and investigate the relevant bodily-perceptual comportments and habits. Then we could make the same considerations about those potential alternative stabilities, the murder weapon, the objet d'art, and such. The notion of relational strategies is thus useful for highlighting the learned conceptions and bodily-perceptual habits involved in using a technology for a particular purpose in a particular context.

Where the hammer provides an example of embodiment relations, we can also consider the relational strategies involved in hermeneutic relations. In such cases, a "hermeneutic strategy" enables a user to apprehend the meaning of a technology's readout in terms of a particular stability, and do so in a perceptual gestalt (e.g., Rosenberger 2011a; 2011b; 2013a). Such a hermeneutic strategy would involve the interpretive framework and the perceptual training that enables a user to read off a particular stability. For example, if a person can at first see only one stability of the Necker cube illusion, they may become able to see another stability after a friend relays an alternative hermeneutic strategy. As Ihde explains, "In a hermeneutic strategy, stories and names are used to create an immediate noetic context. . . . The story creates a condition that immediately sediments the perceptual possibility" (1986, 88). In the example of the doctors arguing over the implications of an fMRI image readout, each doctor brings a different hermeneutic strategy to bear on the image, thus enabling a different stable perceptual interpretation.

Kyle Powys Whyte reviews a series of postphenomenological case studies and examples that utilize the notion of multistability, and he develops methodological recommendations (forthcoming). According to Whyte, examples can be categorized with regard the kind of technological target that is analyzed *as* multistable. Whyte observes that in one category of postphenomenological case studies, it is a wide set of practices that is investigated in terms of its multistability. An example is Ihde's recent analysis of archery in which he considers different styles of bow as they have been used in different cultures throughout history, with the English longbow and the Mongolian horse bow as examples of different stabilities (2009, 16–19). In a second category, it is a single device (or kind of device, or set of devices) that is analyzed with regard to its multiple potential stabilities. This is the case in the example of the Paro robot finding different meanings in different cultures; it is the same device which supports different stable meaningful relations. It is also the case in the example above of the failing computer in which different possible relationships to the same device are cross-analyzed. Whyte uses the term "pivot" to refer to each of these different forms of multistability, and recommends that postphenomenologists remain explicit about the nature of the pivots utilized in their analyses.

## POSTPHENOMENOLOGICAL METHODOLOGY

As said in the introduction to this chapter, an increasing number of scholars have started to use the postphenomenological approach as a method for investigating science and technology. Postphenomenology can in fact be seen as a form of "empirical philosophy." This name may sound like an oxymoron. Philosophy is not an empirical field, after all. Yet, an empirical orientation can be a full-blown part of doing philosophy, not as a positivist basis of philosophical "knowledge," but as a concrete starting point for philosophical reflection (cf. Achterhuis, 2001; Mol, 2002).

As "empirical philosophy," postphenomenology does not base itself on the philosophical tradition and on conceptual analysis only, but also on the study of actual technological practices and artifacts. In doing so, it does not merely "apply" philosophical analyses to science and technology, but it investigates the implications of such practices and artifacts for philosophical conceptualizations. Rather than analyzing the imaging devices on the Mars orbiter in terms of existing theories of perception or knowledge development, it investigates how this device establishes a new kind of sensorial relationship between scientists and the planet Mars, which is a new basis for knowledge development. And rather than assessing obstetric ultrasound in terms of pre-given normative frameworks it investigates how this technology urges us to conceptualize the way technologies help to shape normative frameworks, and to make room for a moral significance of technology in ethical theory.

This empirical-philosophical approach does not follow a strict postphenomenological methodology. The approach is too diverse and too context-sensitive for that. Still, postphenomenological studies have some characteristic elements in common. First of all, they typically focus on *understanding the roles that technologies play in the relations between humans and world*, and on analyzing the implications of these roles. It is on the basis of these relations that technologies have an impact on human beings, and on culture and society. Moreover, it is in the context of these relations that technologies are used and interpreted.

This focus on human-technology relations implies, second, that *postphenomenological studies always include empirical work as a basis for philosophical reflection*. In order to understand human-technology relations, an empirical account is required of the role actual technologies play in human experiences and practices. Such an account can be developed on the basis of empirical work by others, from self-conducted studies, or from an analysis of first-person experiences that involve specific technologies. The purpose of this empirical work is not to develop an accurate description of specific technologies, but to investigate the character of the various dimensions of the relations between humans and these technologies, and their impact on human practices and experiences.

Third, *postphenomenological studies typically investigate how, in the relations that arise around a technology, a specific "world" is constituted, as well as a specific "subject."* As discussed above, by organizing a relation between human beings and their environment, technologies constitute a specific "objectivity" of the experienced world as well as a specific "subjectivity" of the person dealing with the technology. The Mars Orbiter Camera, for instance, lets the planet Mars be "real" in a very specific way, and enacts a very specific type of "observer," just like MRI scanners constitute the reality of the brain in a very specific way, as well as the ways in which psychologists or neurologists can study and understand it.

Fourth, on the basis of these three *elements, postphenomenological studies typically make a conceptual analysis of the implications of technologies* for one or more specific dimensions of human-world relations—which can be epistemological, political, aesthetic, ethical, metaphysical, et cetera. The central question then is how technologies help to shape knowledge, politics, aesthetic judgments, normative ideas, religious experiences, et cetera.

Just like congenial approaches in STS, like Actor-Network Theory, then, postphenomenological research does not have a strict methodology. Rather, it embodies a specific way of investigating technologies, an approach to technology that combines an empirical openness for the details of human-technology relations with phenomenological conceptualization. While ANT focuses on following the *actors* around specific technologies or scientific practices or results, postphenomenology aims to analyze the networks of *relations* around technologies.

## CASE STUDIES

An essential aspect of the postphenomenological perspective is its focus on case studies of concrete human-technology relations to technologies. This case study approach reflects postphenomenology's commitment to the "empirical turn" and its pragmatic antifoundationalism. (This focus on concrete cases makes postphenomenology a part of the empirical turn in the philosophy of technology [see, Kroes and Meijers, 2000; Achterhuis, 2001].) That is, in contrast to totalizing accounts that make claims about how all technologies must be, postphenomenological claims are posed and evaluated within the contexts of studies of particular relations between humans and technologies.

Postphenomenological case studies play a dual role for this philosophical perspective. First, they instantiate the concepts and commitments of the postphenomenological framework. If postphenomenology proves its value through practical contributions to contemporary interdisciplinary conversations on design, scientific investigation, policy, and so on, then case studies will be a primary vehicle for demonstrating this value. In pragmatic spirit, Ihde writes, "The test, however, should lie in outcomes—what produces the relatively better analysis, interpretation, or critique?" (2003, 136). Second, case studies are at the same time the laboratories within which postphenomenological ideas are interrogated and refined. By applying postphenomenological concepts to concrete instances of human-technology relations, the advantages, disadvantages, limits, and places of potential expansion and enrichment can be identified. As a final exercise for this "field guide," let's review a few of the more representative and influential cases conducted through this philosophical perspective.

### Imaging Technologies

Perhaps the most expansive set of studies in postphenomenology regards the use of imaging technologies in scientific and medical practice (e.g., Ihde, 1998; Hasse, 2008; Verbeek, 2007; Carusi, 2009; Rosenberger, 2011a;

2011b; 2013b; Verbeek, 2011; Forss, 2012; Friis, 2012a; Friis, 2012b; Carusi & Hoel, 2014; Hasse, 2014; see also Riis on architectural façades, 2011).

While Ihde explores examples of image-making processes in many of his works, his most systematic treatment appears in his 1998 book, *Expanding Hermeneutics: Visualism in Science*, the last third of which contains a "mini-monograph" on the phenomenology of laboratory imaging practice. Broadly speaking, a user's relationship to an image in science can be understood as a hermeneutic relation. That is, a user can be understood to share a reading-style relationship with an image. An imaging device transforms an otherwise imperceptible aspect of the world into a readable form—an image. As the user looks directly at and interprets the image readout, she or he receives a transformed experience of the world. The image is interpreted by the user in much the same way a person apprehends written language; if the user knows how to interpret the image—if she or he knows "how to read" it—then much of the context and meaning appears at once in a perceptual gestalt. According to Ihde, it is exactly the capacity to provide information in the form of a gestalt that makes images so crucial to contemporary scientific practice. He writes, "The role of repeatable, Gestalt patterns . . . is the epistemological product of this part of the quest for knowledge" (Ihde, 1998, 171).

This series of postphenomenological insights into the hermeneutics of imaging technologies serves to articulate several aspects of scientific practice that typically go underexamined in more conventional accounts. Rather than understand images simply as a representation of the world, or perhaps as a data set to be squared away with theory, the postphenomenological perspective reveals the essential relational dimension of these practices, as a human user encounters a technologically mediated world. Two crucial points become apparent.

First, scientific images are clearly *not* a simple encounter with the world itself. Images and imaging technologies are better understood *as* technologies, that is, *as* transformative mediators of human experience. This perspective thus highlights the role of an imaging technology in transforming an otherwise impossible-to-see object of study into a form perceivable by the human body: far away objects are brought near; microscopic objects are enlarged; hyperfast objects are slowed to a frame; internal bodily processes are opened to view; etc. According to Ihde, the "intentionality" of imaging technologies can even be seen to lead the course of scientific development. Traditional accounts in the philosophy of science tend to understand scientific development to be exclusively (or at least primarily) piloted by the development of theory, with experimentation tagging along in the role of a theory's evaluator. In contrast, Ihde contends that science is more primarily dragged forward through the development of instruments, following "instrumental trajectories" in which those devices are more and more refined.

Second, the postphenomenological perspective spotlights the role of human bodily perceptual experience in scientific imaging practice. When a scientist encounters an object of study prepared in the form of an image, she or he brings a trained form of vision to the apprehension of the image's content. That is, as an expert, as someone deeply familiar with the practice of interpreting specific images, the scientist brings a set of visual habits to the perception of an image. The image interpretation process thus does not reduce to a kind of cognitive translation of image content. Instead, through learned and sedimented perceptual habits, much of an image's content is apprehended in the context of a gestalt. The mediation process can thus be understood to coshape both the scientific observer and the world under observation. Much of what it means to do this form of scientific investigation, and what it means to be an object of study, is constituted through the mediation of imaging technologies.

From these basic points about the hermeneutics of scientific imaging, the school of postphenomenology has produced an array of concrete case studies. Let's focus here on two issues which these studies have explored: the multistability of images in science, and the moral implications of imaging practices.

Hasse's studies on image-reading training practices in physics are a paradigmatic example of a positive use of the notion of multistability. She develops a postphenomenological account of the cultural specificities involved in teaching physics students how to interpret images in the laboratory (e.g., Hasse, 2008; 2014). Central to her analysis is a conception of imaging technologies as multistable mediators of human experience; laboratory images are capable of supporting multiple stable hermeneutic relations. Hasse investigates these multiple potential stabilities in concrete terms, comparing, for example, different image reading practices in different countries, and the different meanings the same images convey to physicists of different generations.

Rosenberger's studies of the multistability of scientific imaging practice share a similar line of thinking. But he considers the ways that the multistability of images facilitates ongoing scientific debates (e.g., Rosenberger, 2011a; 2011b; 2013b). Like Hasse, he argues that it is productive to conceive of laboratory images as multistable mediators capable of facilitating multiple coherent hermeneutic relations. In his studies it is different sides of ongoing debates that supply the hermeneutic strategies that make possible the different stabilities of the image. That is, each side of a debate supplies an interpretive framework that enables the image to be perceived in terms of their own position. Such a conception, Rosenberger claims, highlights the image's status *as* a technology (rather than simply as a representation of reality, a representation about which one side of the debate is right and the other wrong). And it encourages reflexivity with regard to the imaging processes that transform an otherwise imperceptible object of study into a form we can see with the human eye. His work includes case studies into flash freezing techniques used in the study of neurotransmission, and satellite-mounted cameras used in the exploration of Mars.

Verbeek investigates the mediation of ultrasound technology (2008b; 2011). Through a hermeneutic relation to the device, a sonogram provides transformed access to a developing fetus in the form of a live-feed picture on a screen. As such, Verbeek presents ultrasound as a guiding example of the non-neutral manner in which mediating technologies co-shape the actors involved in its use. That is, Verbeek shows how this mediating technology crucially and multiply informs how the parents, the fetus, the fetus's environment, and the medical context are all constituted within the moral decision-making circumstances surrounding pregnancy. He observes, for example, that an ultrasound can have an effect of isolating the fetus, constituting it as an individual, and constituting the mother as the environment within which that individual develops. And more, the mediation of the ultrasound constitutes the fetus *as* a patient, and the course of the pregnancy is correspondingly constituted as a condition which requires the supervision of medical professionals. Verbeek emphasizes that parents are additionally constituted by the ultrasounds as decision makers, forced to contend with morally fraught information regarding risk factors, potential birth defects, and the probabilities regarding conditions such as Down syndrome. The effects of this mediation on contemporary pregnancy are such that even parents that decide against getting an ultrasound are placed in the moral decision-making role of opting out. All this serves as a dramatic illustration of the postphenomenological claim, most strongly expounded and developed by Verbeek, that technological mediation does not reduce to the mere usage of innocent instruments for human purposes, but is instead the non-neutral context through which human actors and the world itself are constituted.

## Implantations

The "posthumanist" discussion tends to make bold predictions about a future that will involve bodies severely altered by technologies, hyper-intelligent machines, and minds downloaded into computer networks. Independent of the most science-fiction-oriented claims, it seems clear that these themes have wide-ranging implications for the philosophy of technology in general, and postphenomenology in particular. For his part, Ihde has tended to be dismissive of at least the most extreme posthumanists (see esp., 2008). But he comes tantalizingly close to fully confronting these issues in his recent essay, "Aging—I Don't Want to be a Cyborg" (Ihde, 2008, ch. 3). Here he reflects on the personal experience of several recent surgeries.

Ihde's central claim is that implanted "cyborg" technologies ultimately remain examples of embodiment relations; implanted technologies are incorporated into a user's bodily encounter with the world; are experienced with more or less transparency; involve a magnification of certain aspects of experience and a reduction of others; and, as all human-technology relations, entail tradeoffs. As Ihde puts it, our thinking should remain in the context of "what I have called *embodiment relations*, which were experienced uses of technologies which remain detachable, but which in use are quasi-transparent, and not technologies taken literally into or inside my body. Yet, I will maintain that the desire remains applicable to *cyborg technologies*" (2008, 32). Ihde considers personal examples such as tooth crowns and his heart stent, describes the experience of their embodiment, and reviews the tradeoffs associated with these non-neutral devices.

Ihde dramatizes these issues with a detailed review of his own 2007 triple bypass surgery, the aftermath of which included dangling electrodes, a leg drain, and an external pacemaker. He lists complications possible for those going through such a procedure, such as disorientation, pain, congestion, and of course the potential for stroke. After being discharged from the hospital, and leaving behind the pacemaker and various monitoring technologies, many after-effects remain, including obligations to wear an event monitor and take medications, medications which themselves carry along new risks for side effects. Though an example of an implanted technology—or more specifically in this case, an elaborate patch of an internal organ—Ihde notes that despite the internal nature of this procedure, we can see in this example all of the expected issues arising from the non-neutrality of human-technology relations.

If Ihde's account can be understood as an attempt to apply the standard postphenomenological framework to the issue of implanted technologies, then there also exists a developing literature which explores the possibility that the notion of implantation presents a challenge to that framework (e.g., Welton, 2006; Verbeek, 2008a; De Preester, 2011; Verbeek, 2011; Besmer, 2012; De Preester, 2012; Dalibert, 2014; Rosenberger, forthcoming). It may be the case that to capture the important aspects of the experience and effects of implanted devices, we must go beyond the concepts of mediation, embodiment, transparency, and tradeoffs. Verbeek's notion of "cyborg relations," reviewed in detail above, is the path-clearing idea in this new line of thinking. This notion is offered specifically to challenge the adequacy of the standard postphenomenological account for addressing examples in which the distinction between the "human" and the "technology" of the human-technology relation is no longer clear.

Another central contribution to this line of thought is made by Helena De Preester, bringing together concerns from both phenomenology and cognitive science (De Preester, 2011; 2012). De Preester argues that the important distinction is not between external devices and prosthetic/implanted devices; it should be between devices that extend our capacities and devices that are incorporated into our bodies—but bodily incorporation should be understood in a phenomenological, and not strictly physical, sense. On what shall we base the distinction? It cannot be based on relative transparency, "since tool use shares this characteristic of withdraw into the sensorimotor body" (De Preester, 2011, 124). Technologies that extend our bodily awareness can be embodied with a high degree of transparency. De Preester settles on the notion of "bodily ownership" as the decisive factor: "Tool use induces changes in motor and sensory capacities, but not in body ownership. True incorporation, in contrast, involves changes in the feeling of body ownership itself" (2011, 123). From here, De Preester explores the different ways in which the distinction between body extending and body incorporated devices applies to prosthetic limb technologies, perception transforming technologies, and extended cognition technologies.

Kirk Besmer's postphenomenological account of the experience of cochlear implant devices has emerged as the go-to case study in this line of thought (2012). Such devices function by first picking up sound through an external microphone, translating that sound through a speech processor, and then sending a radio signal to stimulate the auditory nerve via a permanent implant surgically attached to the cochlea. Besmer reviews the complex training process involved in growing accustomed to the implant, learning to distinguish foreground and background sounds, and ascertaining speech patterns. He explains that, "Activating the implant is a moment in which wearers are torn from silence and thrown into a world of noise, which in phenomenological terms is a re-worlding" (Besmer, 2012, 304). Besmer suggests that cochlear implants are an example of cyborg relations, noting that, "This is the 'merging' of human and machine that Peter-Paul Verbeek talks about" (2012, 306). But he goes further to suggest that these devices constitute their own form of intentionality, one that calls together the bodily characteristics of embodiment relations, and also the translational characteristics of hermeneutic relations, which he calls "cyborg intentionality."

## Cell Phones

Considering just how normal it has become for cell phones and smartphones to be carried around throughout the day, it is perhaps no surprise that postphenomenologists have devoted considerable attention to these devices. For example, in an expansion of the notion of alterity relations through an analysis of the works of Emmanuel Levinas, Galit Wellner has developed a conception of smartphone screens that emphasizes their role as a "quasi-face" point of interaction (2014b). Evan Selinger considers the multistability of cell phones in their use as part of microcredit loan programs in Bangladesh

(2009). And Ingrid Richardson has reflected on the changing conceptions of presence and location that are evolving with the increased use of mobile phones and computing (e.g., Richardson, 2005; Richardson and Wilken, 2011).

In Rosenberger's work on cell phone usage, he develops a postphenomenological account of the experience of phone conversation (e.g., Rosenberger, 2012; 2013c; 2014b). His contention is that, due in part to a user's deep familiarity with the phone, the content of phone conversation occupies much of the space of the user's mind. This experience can be articulated in detail through the use of postphenomenological tools. In this account, phone usage should be understood as an embodiment relation, one that transforms a user's capacity to communicate with others over a distance. As a user becomes accustomed to the device, the phone itself becomes highly transparent in the user's hand, even as the device renders such extensive transformations to the user's capacities, and even as the user continues to actively hold the phone beside her or his face.

Rosenberger argues further that typical phone usage is also highly characterized by sedimentation and field composition. (Indeed the notions of "field composition" and "sedimentation" were first developed within this work in part as an attempt to expand postphenomenology so that it could capture the experience of the phone and related technologies.) To say that phone usage is highly characterized by field composition is to say that the phone significantly reorganizes a user's overall field of awareness. The specific organization inclined by the phone is one in which the content of the conversation and the mediated presence of the interlocutor stand forward and "fill up" much of what the user is experiencing. This is why, for example, if a person stands alone in a room and holds a conversation over the phone, she or he may not "see" the objects of the room even though this person's eyes remain open and looking forward. And more, Rosenberger claims that these features of phone usage—the significant field constitution, the high level of transparency of the device itself—all occur in a deeply sedimented fashion. That is, these features often snap into place with considerable immediacy and stubbornness due to the user's individual long-developed bodily perceptual habits regarding the phone.

An intra-postphenomenological debate has recently emerged over the topic of cell phones and driving between Wellner and Rosenberger. Rosenberger has become an outspoken critic of the practice of driving while using the phone, and advocates greater regulations (e.g., Rosenberger, 2012; 2013c; 2014b). His argument consists of an interpretation of the preponderance of empirical data that show both handheld and hands-free phone usage to accompany a dangerous drop in driving performance. He claims that this drop is due to the features of the experience of the phone discussed above; the driver's deeply sedimented relationship to the phone inclines a field composition directed mostly upon the conversational content and the telepresence of the interlocutor. That is, according to Rosenberger, due to the deeply-entrenched habits of the phone, a driver's mind is at times dangerously pulled away from the road and toward the phone conversation with a force akin to a bad habit. Wellner remains unconvinced. She argues that users can develop ways to successfully split their attention between two tasks, what she calls "multi-attention," and thus that increased regulation is unnecessary (Wellner, 2014a). Their debate has dragged a number of postphenomenologists into the discussion (see the 2014 special issue of *Techné* on this topic), and addresses not only the concrete details of driver experience and traffic policy, but also the nature of human attention, and the relation between habituation and multistability.

## REFERENCES

Achterhuis, H. (ed.). (2001). *American Philosophy of Technology: The Empirical Turn*. Bloomington: Indiana University Press.

Besmer, K. (2012). "Embodying a Translation Technology: The Cochlear Implant and Cyborg Intentionality." *Techné: Research in Philosophy and Technology* 16 (3): 296–316.

Bottenberg, F. (2014). "Searching for Alterity: What Can We Learn From Interviewing Humanoid Robots?" In: *Postphenomeonlogical Investigations: Essays in Human-Technology Relations* (this volume).

Carusi, A. (2009). "Philosophy engines: Technology and reading/writing/thinkingphilosophy." *Discourse* 8 (3) http://www.prs.heacademy.ac.uk/view.html/PrsDiscourseArticles/69

Carusi, A. and A. S. Hoel. (2014). "Toward A New Ontology of Scientific Vision." In Coopmans, Vertesi, Lynch, and Woolgar (eds.), *Representation In Scientific Practice Revisited.* Cambridge, MA: MIT Press 201–221.

Dalibert, L. (2014). "Posthumanism and Somatechnologies: Exploring the Intimate Relations Between Humans and Technologies." Doctoral dissertation, University of Twente.

De Preester, H. (2011). "Technology and the Body: The (Im)Possibilities of Re-embodiment." *Foundations of Science* 16: 119–137.

De Preester, H. (2012). "Technology and the Myth of the Natural Man." *Foundations of Science* 17: 385–390.

Eason, R. (2003). "Hypertext: Rortean links between Ihde and Haraway." In D. Ihde and E. Selinger (eds.), *Chasing Technoscience: Matrix for Materiality*. Bloomington: Indiana University Press, 167–181.

Feenberg, A. (2000). "From Essentialism to Constructivism: Philosophy of Technology at the Crossroads." In E. Higgs, D. Strong, and A. Light (eds.), *Technology and the Good Life*. Chicago: Univ. of Chicago Press, 294–315.

Forss, A. (2005). *At the Margins of Health and Normality: Women's Encounters with Biomedical Technology in the Realm of Cervical Cancer Screening*. Stockholm: Repo Print AB.

Forss, A. (2012). "Cells and the (Imaginary) Patient: The Multistable Practitioner–Technology–Cell Interface in the Cytology Laboratory." *Medicine, Health Care, & Philosophy* 15: 295–308.

Friis, J. K. B. O. (2012a). "Interpreting the Visual." *Philosophy & Technology*. 25: 249–253.

Friis, J. K. B. O. (2012b). "Perception: Embodiment and Beyond." *Foundations of Science* 17: 363–367.

Gurwitsch, A. (1964). *Field of Consciousness*. Pittsburgh: Duquesne University Press.

Hasse, C. (2008). "Postphenomenology: Learning Cultural Perception in Science." *Human Studies* 31 (1), 43–61.

Hasse, C. (2013). “Artifacts That Talk: Mediating Technologies As Multistable Signs and Tools.” *Subjectivity*. 6: 79–100.
Hasse, C. (2014). *The Anthropology of Learning: On Nested Fictions and Cultural Ecologies*. Dordrecht: Springer.
Heidegger, M. (1996). *Being and Time*, trans. J. Stambaugh. Albany: SUNY Press.
Husserl, E. (1973). *Logical Investigations*. London: Routledge.
Ihde, D. (1986). *Experimental Pheneomenology, second printing*. Albany: SUNY Press.
Ihde, D. (1990). *Technology and the Lifeworld*. Bloomington: Indiana University Press.
Ihde, D. (1993). *Postphenomenology: Essays in the postmodern context*. Evanston, IL: Northwestern University Press.
Ihde, D. (1998). *Expanding Hermeneutics: Visualism in Science*. Evanston: Northwestern University Press.
Ihde, D. (1999). “Technology and Prognostic Predicaments.” *AI & Society*. 13: 44–51.
Ihde, D. (2003). “If Phenomenology is an Albatross, Is Post-phenomenology Possible?” In D. Ihde and E. Selinger (eds.), *Chasing Technoscience*. Bloomington: Indiana University Press, 131–144.
Ihde, D. (2007). *Listening and Voice,* 2nd edition. Albany: SUNY Press.
Ihde, D. (2008). *Ironic Technics*. New York: Automatic Press/VIP.
Ihde, D. (2009). *Postphenomenology and Technoscience: The Peking University Lectures*. Albany: SUNY Press.
Ihde, D. (2012). *Experimental Phenomenology: Multistabilities, Second Edition*. Albany: SUNY Press.
Irwin, S. (2006). “Technological Other/Quasi Other: Reflection on Lived Experience.” *Human Studies* 28: 453–467.
Kroes, P., and A. Meijers (eds.). (2000). *The Empirical Turn in the Philosophy of Technology*. Amsterdam: JAI.
Latour, B. (1993). *We Have Never Been Modern*. Trans. C. Porter. Cambridge, MA: Harvard University Press.
Merleau-Ponty, M. (1962). *Phenomenology of Perception*. Trans. Colin Smith. London: Routledge & Kegan Paul.
Richardson, I. (2005). “Mobile Technosoma: Some Phenomenological Reflections on Itinerant Media Devices.” *Fibreculture Journal* 6. http://six.fibreculturejournal.org/fcj-032-mobile-technosoma-some-phenomenological-reflections-on-itinerant-media-devices/
Richardson, I., and R. Wilken. (2011). “Parerga of the Third Screen: Mobile Media, Place, Presence.” In G. Goggin and R. Wilken (eds.), *Mobile Technology and Place*. New York: Routledge.
Riis, S. (2011). “Dwelling In-Between Walls: The Architectural Surround.” *Foundations of Science* 16: 285–301.
Rosenberger, R. (2008). “Perceiving Other Planets: Bodily Experience, Interpretation, and the Mars Orbiter Camera.” *Human Studies* 31 (1).
Rosenberger, R. (2009). “The Sudden Experience of the Computer.” *AI & Society*. 24: 173–180.
Rosenberger, R. (2011a). “A Case Study in the Applied Philosophy of Imaging: The Synaptic Vesicle Debate.” *Science, Technology, & Human Values* 36 (6): 6–32.
Rosenberger, R. (2011b). “A Phenomenology of Image Use in Science: Multistability and the Debate Over Martian Gully Deposits.” *Techné: Research in Philosophy and Technology* 15 (2): 156–169.
Rosenberger, R. (2012). “Embodied Technology and the Problem of Using the Phone While Driving.” *Phenomenology and the Cognitive Sciences* 11 (1): 79–94.
Rosenberger, R. (2013a). “Mediating Mars: Perceptual Experience and Scientific Imaging Technologies.” *Foundations of Science* 18: 75–91.
Rosenberger, R. (2013b). “The Importance of Generalized Bodily Habits for a Future World of Ubiquitous Computing.” *AI & Society* 28: 289–296.
Rosenberger, R. (2013c). “The Problem with Hands-Free Dashboard Cell Phones.” *Communications of the ACM* 56 (4): 38–40.

Rosenberger, R. (2014a). "Multistability and the Agency of Mundane Artifacts: From Speed Bumps to Subway Benches." *Human Studies* 37: 369–392.
Rosenberger, R. (2014b). "The Phenomenological Case for Stricter Regulation of Cell Phones and Driving." *Techné: Research in Philosophy & Technology* 18(1/2): 20–47.
Rosenberger, R. (forthcoming). "Postphenomenology: What's New? What's Next?" In J. K. B. O. Friis and R. P. Crease (eds.), *Technoscience and Postpheomenology: The Manhattan Papers*. Lanham, MD: Lexington Books.
Scharff, R. C. (2006). "Ihde's Albatross: Sticking to a "Phenomenology" of Technoscientific Experience." In E. Selinger (ed.), *Postphenomenology: A Critical Companion to Ihde*. Albany: State University of New York Press, 131–144.
Selinger, E. (ed.). (2006). *Postphenomenology: A Critical Companion to Ihde*. Albany, NY: State University of New York Press.
Selinger, E. (2009). "Towards a Reflexive Framework For Development: Technology Transfer After The Empirical Turn." *Synthese* 168: 377–403.
Van Den Eede, Y. (2011). In Between Us: On the Transparency and Opacity of Technological Mediation." *Foundations of Science* 16: 139–159.
Verbeek, P.-P. (2005). *What Things Do: Philosophical Reflections on Technology, Agency, and Design*. University Park: Penn State University Press.
Verbeek, P.-P. (2007). "Beyond the Human Eye: Technological Mediation and Posthuman Visions." In P. Kockelkoren (ed.), *Mediated Vision*. Rotterdam: Veenman Publishers, 43–53.
Verbeek, P.-P. (2008a). "Cyborg Intentionality: Rethinking the Phenomenology of Human-Technology Relations." *Phenomenology and the Cognitive Sciences* 7(3): 387–395.
Verbeek, P.-P. (2008b). "Obstetric Ultrasound and the Technological Mediation of Morality: A Postphenomenological Analysis." *Human Studies* 31: 11–26.
Verbeek, P.-P. (2011). *Moralizing Technology: Understanding and Designing the Morality of Things*. Chicago: Chicago University Press.
Wellner, G. (2014a). "Multi-Attention and the Horcrux Logic: Justifications for Talking on the Phone While Driving." *Techné: Research in Philosophy & Technology* 18 (1/2): 48–73.
Wellner, G. (2014b). "The Quasi-Face of the Cell Phone: Rethinking Alterity and Screens." *Human Studies* 37: 299–316.
Welton, D. (2006). "Body and Machines." In E. Selinger (ed.), *Postphenomenology: A Critical Companion to Ihde*. Albany: SUNY Press, 197–209.
Whyte, K. P. (forthcoming). "What is Multistability? A Theory of the Keystone Concept of Postphenomenological Research." In J. K. B. O. Friis and R. P. Crease (eds.), *Technoscience and Postphenomenology: The Manhattan Papers*. Lanham, MD: Lexington Books.

# 2

# Postphenomenological Theories

*Chapter Two*

# Why Postphenomenology Needs a Metaphysics

Lenore Langsdorf

> I don't think phenomenology returns to the given. . . . Once you have been able to discover the possible multi-stability of a phenomenon, you can never return to it as a given.
>
> —Don Ihde, *Chasing Technoscience* (2003, 127)

> There are, from Heraclitus to Bergson, philosophies, metaphysics, of change. One is grateful to them for keeping alive a sense of what classic, orthodox philosophies have whisked out of sight. But the philosophies of flux also indicate the intensity of the craving for the sure and fixed.
>
> —John Dewey, *Experience and Nature* (*Later Works* 1) (1925, 49)

> The many become one and are increased by one.
>
> —Alfred North Whitehead, *Process and Reality* (1978/1929, 21)

There are two responses to this title's implicit question. One is brief: All of our investigations, indeed all of our mundane activities, are based on metaphysical assumptions and convictions, most of which go unnoticed. Unexpected results and scientific revolutions sometimes make those thematic. Phenomenology's basic interest in making thematic what was unnoticed supports investigation of this implicit metaphysics.

The second response is longer, for it begins with Dewey's inclusion among the philosophers who provide "philosophies, metaphysics, of change" to philosophy's history. His event ontology envisioned the world as ordered occurrences of organisms' interaction with their environment. His naturalistic metaphysics envisioned nature, and the universe itself, as composed of those interactive events in an ongoing process of "potential consequences" which, as they are integrated, "form the very nature and essence of a thing, its defining, identifying and distinguishing form" (Ihde, 1976, 143).

That aspect of postphenomenology's pragmatic heritage is evident in Don Ihde's discovery of the multi-stability of phenomena, and the possibility that that character is a manifestation of the multistability of "the things themselves." This is an ontological topic; as Ihde noted, his *Listening and Voice* was "intended as a prolegomena to an ontology of listening" (1976, ix), and expanded, as he integrated the visual realm with the aural/oral, into a "phenomenological ontology" (2003, 134), "relativistic ontology" (2006, 275), and, most recently, an "interrelational ontology" (2009, 44). The later terms term imply issues of identity and change, mind and matter, possibility and necessity. These issues go beyond description of what is and into questions of how multistable things come to be; in more phenomenological terms, how that multi-stability is constituted. These are topics for metaphysical inquiry.

The pervasive historical dominance in Euro-American philosophy of Plato's promotion of stable permanence (Forms) over change (things), together with the focus on the immutable in Aristotelian metaphysics, contribute to the avoidance of metaphysical study in modern philosophy. Yet the multistability of postphenomenology's "things" encourages a broadened interest in how "being as such" encompasses the becoming of that being; in Dewey's words, in the possibilities of a "metaphysics, of change" despite "the craving for the sure and fixed" (1925, 49).

## POSTPHENOMENOLOGY

> Postphenomenology is a hybrid phenomenology. On the one hand, it recognizes the role of pragmatism in the overcoming of early modern epistemology and metaphysics. . . .On the other side, it sees in the history of phenomenology a rigorous style of analysis through the use of variational theory, the deeper phenomenological understanding of embodiment and human active bodily perception, and a dynamic understanding of a lifeworld as a fruitful enrichment of pragmatism.
>
> —Don Ihde, *Postphenomenology and Technoscience* (2009, 23)

The hybridity of postphenomenology reorients it from Husserl's predominant orientation toward the "the things themselves" as they appear, available for human investigation in terms of conceptual entities and static essences. It enables a turn, instead, toward pragmatism's orientation, most marked in John Dewey's work, toward concrete cultural (historical, social, and political) events as occasions of the interaction of environments and organisms. Further, the focus on technologies as practices instigates attention to the "how" of their interrelations and the positive and negative value that emerges in those practices.

Ihde recognizes that the "organism/environment notion of interactionism" of a "pragmatically bonded phenomenology" aligns him with Dewey's "concrete and practice-oriented direction" (2003, 136). However, he notes that other postphenomenological researchers—he names Donna Haraway and Bruno Latour—are more aligned with Alfred North Whitehead's work: "Whitehead's 'process'-oriented philosophy clearly resonates with all of the contemporary praxis versions of technoscience studies." However, he continues, Whitehead's "special vocabulary of neologisms put [him] off" (Ihde, 2003, 135).

In what follows I propose that postphenomenology's "interrelational ontology" needs metaphysics, and that the metaphysics it needs is a process metaphysics that provides a theory of how interrelation produces novel entities. Neologisms are unnecessary if that theory is articulated, as in ethnomethodology; namely, from observation of actual practices, rather than (as in Whitehead's explication) at a distance from the fieldwork and laboratory in and by which actual multistable objects are discovered. The metaphysics I have in mind is summarized in Bruno Latour's reliance upon a metaphysics that is "a very concrete practice… a mixture of ethnomethodology and ontology" that enables "metaphysicians to monitor the experiment in which the world makes itself" (Latour, 2003, 18). This is a metaphysics that refuses philosophers' persistent and intense "craving for the sure and fixed" against which Dewey inveighed. Advocating for it begins in thematizing the ontology that postphenomenology affirms, in contrast to its Husserlian predecessor, together with elucidating ethnomethodology's practices.

## ONTOLOGY

> In both pragmatism and phenomenology, one can discern what could be called an interrelational ontology. By this I mean that the human experiencer is to be found ontologically related to an environment or world, but the interrelation is such that both are transformed within this relationality.
>
> —Don Ihde, *Postphenomenology and Technoscience* (2009, 23)

One rather standard response to the question of the difference between metaphysics and ontology (granted that the terms are often used interchangeably) is that ontology deals with entities; with *what* is in the environment, and metaphysics deals (along with other topics) with *how* those entities have their being. If those definitions are accepted, we can see why the basic interest of traditional phenomenology was ontology, with a strong emphasis on epistemology. The goal of phenomenology as a "rigorous science" was devising and using a method for discovering what humans could know, with certainty, about the entities correlative with human noetic activity.

An ontology tells us what is in our environment. Husserl, informed and thus forewarned by British empiricism's reduction of things to sensory evidence and mental ideas, as well as Kant's relegation of things themselves to the domain of intrinsically unknowable noumena, limited investigation to what appears to humans and to the constitution of those appearances in relation to human beings. Thus Husserlian phenomenology gives us a relational ontology that is grounded in noetic activity and its noematic correlate; consciousness and its intentional objects. It also gives us a method for investigating the essence of that activity—intentionality—engaged in the constitution of those objects. Yet although phenomenological method revealed the temporal structure of constitution, by way of the presence of past moments and anticipation of future moments within intentional objects' presence, concern for identification of essences suggests static objective correlates to consciousness's activity.

The assault on presence in postmodern philosophy is an understandable, and perhaps inevitable, rebound from that emphasis. The "post" qualifier in postphenomenology may signify both an alignment with postmodernism's deconstructive response to presence and a determination to avoid the "subjectivism" that plagues philosophers' conception of phenomenology. Postphenomenology's inherited method of close observation, however, is a constructive one insofar as it focuses on the ingredients that constitute novel actuality, and its strong focus on technology need not limit that focus to the interrelation of humans and inanimate things (human subjects and their objects).

Insofar as it retains phenomenology's interest in constitution, and especially if it is to incorporate pragmatism's interest in melioration, postphenomenological investigations need attention to the "how" of change, without defaulting to an emphasis on human intentionality that threatens to elevate subjectivity over objectivity. That default can too easily lead to subjectivism—an identification that haunts classical phenomenology, which Don Ihde calls "first phenomenology": "a method and field of study" that "operated like a science" to provide "a statics of experience" (1976, 18–19). He retains that method, although with attention to experience as dynamic, in "second phenomenology," which "builds toward a fundamental ontology of Being" (Ihde, 1976, 18). This shift requires turning from "a latent, presupposed, and dominant visualism in our understanding of experience" (Ihde, 1976, 6). Ihde criticizes this "preference for vision" as "tied to a metaphysics of objects" by the "invention of a perspective" that separated "sense" (explicitly, vision), and sensory objects, from "significance": the "ultimately real was beyond sense"—with the result that "sense was diminished" (1976, 8–9). This separation and diminishing of both sense and significance, he notes, persisted in "classical" empiricism: for Locke, "mundane experience was immediately bypassed for what became empiricist atomism" (Ihde, 1976,

12). Thus, Ihde concludes, "Locke repeated in essential outline the metaphysical division of the thing" (1976, 13).

To undo that division, Ihde's early work accomplished a "deliberate decentering" of "classical" empiricism: "What is called for is an ontology of the auditory" that does not "replace vision as such with listening as such," but enables a "move towards a radically different understanding of experience" (1976, 14–15). The seeds of the focus on technology in postphenomenology are discernable in the early development of this "second phenomenology" as the setting for a "prolegomena to an ontology of listening" (Ihde, 1976, ix). Ihde emphasizes that "the philosopher must listen to the sounds as meaningful"—which is to say, as embodying both "sense" and "significance." He recognizes that "our experience of listening itself is being transformed" by living in a "technological culture" that informs that transformation (Ihde, 1976, 4–5). Yet this ontology is limited to what is experienced by humans, either through utilizing our sensory capacities or through expanding and even supplanting those capacities with technologies that humans devise.

By providing an ontology that departs from "classical" empiricism's understanding of objects and categories of experience, Ihde's development of "second phenomenology" "builds toward a fundamental ontology of Being" (1976, 18). Insofar as that requires consideration of what presents itself in appearances, it prepares for, but does not engage, the very process of experiencing—a process that does not, itself, present itself phenomenologically. In other words, Ihde's "interrelational ontology" (1976, 23) takes us to the threshold of a metaphysical, in contrast to phenomenological, investigation.

Bruno Latour's metaphysical interest takes him beyond that threshold, and informs his assertion that "the phenomenological tradition that he knows best—the one of Merleau-Ponty… is an entirely human-centered account of embodiment" although "it is more helpful than overly rationalistic positions, because of the attention on the lived world" (2003, 16). He acknowledges that he doesn't "see the use of phenomenology" for responding the question that provokes his investigations: "Can we gain access to agencies that are not human-centered?" (Latour, 2003, 17). Given that interest, he "would rehabilitate metaphysical questions that were posed by Whitehead," for whom "the ingredients of the world are accessible to inquiry" (Latour, 2003, 16, 17). It's in fieldwork and laboratory research that he situates inquiry about "the basic categories by which actors build time, space, and agency… categories that should not be fixed in advance" of that research (Latour, 2003, 17).

In sum: Latour's "very concrete" metaphysics calls for a "mixture of ethnomethodology and ontology" (2003, 18) along with semiotics. He considers all three to be the "necessary tools for seeing how some events or actors can remake the world locally" and goes on indicate the particular value of "small doses" of semiotics: "linking it with ethnomethodology" provides a "philosophical tool" that is "effective for bracketing out the subject" and so clears a way for attending to "how the world is built" (Latour, 2003, 22). That bracketing is needed to avoid limitation to human subjects and the risk of adopting, in advance, categories that are specified by phenomenology's limitation to what appears as accessible to human knowledge. Ethnomethodology's focus on practices detects local interactions and interrelations accomplishing world building, and thus, expands the realm of the accessible.

## ETHNOMETHODOLOGY

> Any setting organizes its activities to make its properties as an organized environment of practical activities detectable, countable, recordable, reportable, tell-a-story-about-able, analyzable—in short accountable.
>
> — Harold Garfinkel, *Studies in Ethnomethodology* (1967, 33)

Harold Garfinkel invented the term "ethnomethodology" to name phenomenologically inspired case studies of how members of a society develop methods for organizing and accounting for their environments. As George Psathas characterizes that focus,

> Where others might see "things," "givens," or "facts of life," the ethnomethodologist sees (or attempts to see) *process*: the process through which the perceivedly stable features of socially organized environments are continually created and sustained. (1980, 3; emphasis in original)

This focus required a dramatic shift from traditional sociological theory, which focused on the identifying the institutions, norms, roles, and rules that function to provide a society's structure and within which members of that society act. Treating "social facts" objectively, as "things," was Emile Durkheim's formulation of the positivistic orientation of sociology's founder, August Comte. The latter sought to create a science of the social on the model of the physical sciences: collect and analyze data, preferably in statistical form, following rules for observation that enabled the positing of correlations and patterns that could be replicated. The model for research was Durkheim's study of suicide, which sought to isolate and describe the external forces correlated with that action. The common term for this adherence to objective research was positivism; the goal was compiling data concerning the external forces that govern actors' behavior.

Garfinkel, however, attended to the methods people used to constitute and interpret those "social facts" in order to discern "the formal properties of common-sense activities as a practical organizational accomplishment" (1967, viii). Seeing these properties and the patterns they formed meant focusing on the details internal to constitutive processes; focusing on the "how" rather than the "what." That required fieldwork: bringing the researcher to sites of activities, whether that be street intersections, dinner tables, recreational games, or laboratories. Often, the researcher came not only to observe but also with a strategy for interfering with what was ongoing in order to see how order was (re-)constituted and sustained; that is, on how that particular corner of the world was being built.

Given this interest in how practices internally and interactively generate their objects, the "things themselves" of phenomenology became the "processes themselves" of ethnomethodology. Insofar as a process occurs in a context and has a goal or purpose, it is continuous with both its past and its future. Its independence is constituted by how it takes up aspects of that past and is attracted to prospective ingredients for its future. In effect, activities are moments within a process, both constituted by that process and constituting its future. Thus the context itself, considered as temporally and spatially efficacious, is no longer a separate and stable frame; rather, it is an ingredient in processual, interactive, and interrelational events.

In sum: ethnomethodology investigates methods that members of a collective use to create and sustain the particular ordering of entities that are needed for their world building. That task requires suspending reliance upon the subject (as in semiotics) but also suspending the researchers' own reliance on hypotheses and presuppositions, such as the "fixed in advance categories about objects, subjects, and reality in general" to which Latour objected. Ethnomethodology's sociological heritage provides an orientation toward human practices, rather than toward phenomenology's intentional subjects or postphenomenology's multistable objects. Yet its phenomenological heritage remains in Garfinkel's description of "practical actions as contingent ongoing accomplishments of organized artful practices of everyday life" (1967, 11). Bruno Latour overlooks that heritage in his appreciation of ethnomethodology as a tool for investigating how the world is built in and through those "artful practices." Approaching world building as an expanding compendium of practices and their ongoing accomplishments supports a radical understanding of the processual nature of reality—and thus, goes beyond ontology to metaphysics.

## POSTPHENOMENOLOGICAL METAPHYSICS

> The idea that the immediate traits of distinctively human experience are highly specialized cases of what actually goes on in every actualized event of nature does more that merely deny us the existence of an impassable gulf between physical and psychological subject matter. It authorizes us, as philosophers engaged in forming highly generalized descriptions of nature, to use the traits

> of immediate experience as clews for interpreting our observations of non-human and non-animate nature.
>
> —John Dewey, "The Philosophy of Whitehead" (1941, 647)

> There is order in the plenum. . . in the most ordinary activities of everyday life in their full concreteness, and that means in their ongoingly procedurally enacted coherence . . . it is intractably hard to describe procedurally. . . . The witnessably recurrent details of ordinary everyday practices are constitutive of their own reality.
>
> — Harold Garfinkel, *Ethnomethodology's Program* (2002, 95–97)

> Metaphysics starts in earnest when you grant those you study the same ability to build the basic categories as when you read a treatise by Leibniz or if you read *Process and Reality*. . . . Metaphysics is a very concrete practice. . . . I take it to be a mixture of ethnomethodology and ontology.
> It is the job of metaphysicians to monitor the experiment in which the world makes itself.
>
> —Bruno Latour, *Chasing Technoscience*, (2003, 18)

> When the description fails to include the practice, the metaphysics is inadequate and requires revision. Metaphysics is nothing but the description of the generalities which apply to all the details of practice.
>
> —Alfred North Whitehead, *Process and Reality* (1978/1929, 13)

A postphenomenological metaphysics attends to the "ordinary everyday practices" in and through which mundane as well as scientific objects are constituted. Its heritage from pragmatism, phenomenology, and ethnomethodology enables it to discover, in case studies of "what actually goes on in every actualized event of nature," multistability that is constituted in multiple intentionalities.

Postphenomenology reorients its diverse, but potentially harmonious, heritage by attending to multistability and the reciprocity of roles that humans and nonhumans play in a world of things that are continually renovated. That reorientation is evident in Don Ihde's and Peter-Paul Verbeek's extension of Husserlian phenomenology's turn to "the things themselves" into an investigation of the processual constitution of things. Ihde's variational method demonstrates that "Technologies, by providing a framework for action, do form intentionalities and inclinations within which use patterns take dominant shape" (1990, 141). Verbeek undertakes "a more radical extension of Ihde's concept of 'technological intentionality'" in his investigation of "what things do"; of "how they mediate the intentional relation between humans and world in which each is constituted" (2005, 116). This mediation provides a level of intrinsic efficacy: Things "are not neutral 'intermediaries' between human and world, but mediators; they actively mediate this relation" (2005, 114). "They carry morality," he proposes, "since they help to shape how human beings act" (Verbeek, 2006, 127). From this perspective, Verbeek investigates things not as pregiven entities that assume relations with each other, but as entities that are constituted in their mutual relations. Human beings are what they are by virtue of the way in which they realize their existence in their world, and their world is what it is by the way in which it can manifest itself in the relations humans have to it (2006, 163).

A metaphysics that describes a world of mutual relations is a radically reconstructed metaphysics. It investigates reality as a process of mutually interactive constitution of concrete technologies and human practices; of the things we make and how those things make us. It must begin empirically, as Whitehead insists: "morphological description is replaced by description of dynamic process" within which "any one actual entity involves the actual other actual entities among its components" (1978, 7). It incorporates Dewey's search for "generic traits of experience" and Whitehead's "Metaphysical categories [that] are not dogmatic statements of the obvious; they are tentative formulations of ultimate generalities" and form "a matrix from which true propositions applicable to particular circumstances can be derived" (1978, 8).

Postphenomenology's pragmatic heritage urges awareness of the value of our doings and undergoings, and metaphysical research is no exception. The value of Whitehead's matrix is suggested by Dewey's summary of the value of empirical philosophical methods, which can only be determined by relocating "the conclusions of philosophic inquiry… in the experience out of which they arose, so they may be confirmed or modified." In that way, Dewey continues, "the philosophical results themselves acquire empirical value; they are what they contribute to the common experience of man, instead of being curiosities to be deposited, with appropriate labels, in a metaphysical museum" (1925, 26).

## REFERENCES

Dewey, J. (1925). *Experience and Nature* (Later Works 1). Carbondale: Southern Illinois University Press.

Dewey, J. (1941). "The Philosophy of Whitehead." In P. A. Schilpp (ed.), *The Philosophy of Alfred North Whitehead*, 641–700.

Garfinkel, H. (1967). *Studies in Ethnomethodology*. Englewood Cliffs, NJ: Prentice-Hall.

Garfinkel, H. (2002). *Ethnomethodology's Program: Working Out Durkheim's Aphorism*. A. Warfield Rawls (ed.). Lanham, MD: Rowman & Littlefield.

Ihde, D. (1976/2007). *Listening and Voice*. Athens: Ohio University Press.

Ihde, D. (1986). *Experimental Phenomenology*. Albany: State University of New York Press.

Ihde, D. (1990). *Technology and the Life World*. Bloomington: Indiana University Press.

Ihde, D. (2003). "Interview with Don Ihde." In D. Ihde and E. Selinger (eds.), *Chasing Technoscience*. Bloomington: Indiana University Press.

Ihde, D. (2006). "Forty Years in the Wilderness." In E. Selinger (ed.), *Postphenomenology: A Critical Companion to Ihde*. Albany: State University of New York Press, 267–290.

Ihde, D. (2009). *Postphenomenology and Technoscience: The Peking Lectures*. Albany: State University of New York Press.

Ihde, D. and E. Selinger (eds.). (2003). *Chasing Technoscience*. Bloomington: Indiana University Press.

Latour, B. (2003). "Interview with Bruno Latour." In D. Ihde and E. Selinger (eds.), *Chasing Technoscience*. Bloomington: Indiana University Press.

Psathas, G. (1980). "Approaches to the World of the Study of Everyday Life." *Human Studies* 3(1): 3–17.

Verbeek, P-P. (2005/2000). *What Things Do*. University Park: Pennsylvania State University Press.

Verbeek, P-P. (2006). "The Morality of Things: A Postphenomenological Inquiry." In E. Selinger (ed.)., *Postphenomenology: A Critical Companion to Ihde*. Albany: State University of New York Press.

Whitehead, A. N. (1978/1929). *Process and Reality*, Corrected Edition. David Ray Griffin and Donald W. Sherburne (eds.). New York: The Free Press.

## *Chapter Three*

# What Robotic Re-embodiment Reveals about Virtual Re-embodiment

## *A Note on the Extension Thesis*

Kirk M. Besmer

A central tenet of phenomenology is that one's sense of presence is intimately tied up with one's material body. That is, the sense one has of being "here" or being "there" involves being physically embodied here or there. Recent technologies of telepresence, such as sophisticated video games, immersive virtual environments, and tele-operated robotic systems, complicate this bedrock belief. Robust real-time communicative action in and feedback from a remote environment provide users of these technologies with a sense of presence in the remote environment. Users often feel *as if they are there*, at the remote site, whether it be a virtual or a real location. This altered sense of presence—the sense that one is both "here" and "there"—has been called, "re-embodiment" by philosophers pursuing (broadly conceived) phenomenological analyses of human engagement with technologies. The questions at issue involve the nature of such re-embodiment and its limits.[1]

In this chapter, I will pursue a postphenomenological analysis of two varieties of technological re-embodiment: virtual re-embodiment, which occurs to some degree in video games and more so in immersive virtual environments; and robotic re-embodiment, which occurs, again to varying degrees, with tele-robotic systems, such as tele-surgery and remotely-operated vehicles. I hope to advance towards a postphenomenological account of technological re-embodiment by *arguing against* a tempting but faulty way of understanding such experiences, namely that technologies of telepresence *extend* human embodiment much like ordinary bodily co-located tools, such as the carpenter's hammer or the blind man's cane. My argument will depend on drawing a distinction between technologies that can be assimilated into the body schema and those that must be integrated at the higher level of body image. As a result of my argument, I hope to offer a refinement of what I shall call the "extension thesis," which is current in philosophical studies of technology, as well as cognitive science accounts of "mind." It also appears in popular media analyses of technology. In brief, the extension thesis is the *general* claim that technologies serve as extensions of embodied human perception, agency, and cognition, or alternatively put, that technologies extend human capacities as embodied beings that perceive, act, and think. While it is clear that bodily co-located tools can become technological extensions of carnal embodiment, it is less clear how this applies to technologies of telepresence, such as we find in virtual and robotic re-embodiment. A distinction between two varieties of technological extensions of embodiment is needed.

## VIRTUAL AND ROBOTIC RE-EMBODIMENT

Playing video games is a fascinating experience because one almost instantly and invariably identifies with one's avatar. This makes sense since the avatar functions as the locus of perception and agency in the virtual environment. Like owners taking up residence in a newly-purchased house, game players rapidly adapt to the virtual environment; in doing so, they transpose many of the elements of real-world embodiment and spatiality into the virtual environment. For example, watch players of video games in which the avatar is a third-person representation of a human or human-like figure, and you will see that they often move their real bodies synchronically with their avatar, ducking in the real world, for example, when a projectile is coming at their avatar's head. Moreover, when moving the avatar "body" in virtual space, players quickly develop a sense of its spatial dimensions: they can judge whether or not the avatar "body" will fit through a door, for example. Thus, not only do they identify with the avatar; they also seem to inhabit it.

This anecdotal evidence is supported by empirical research done by social psychologists Jeremy Bailenson and Jim Blascovich who have conducted various studies of human experiences in immersive virtual environments. Although their research is primarily oriented towards issues in social psychology, some of their research is revealing for theories of human embodiment, especially as it pertains to virtual experiences. For example, when controlling an avatar in virtual space, people will respect the personal space of another's avatar; they will do so even when they believe that the avatar is a computer-controlled agent. Likewise, when the avatars of others invade their avatar's personal space by coming too close, they react by moving away. Their studies show that habits of minding socially acceptable interpersonal space established in real, physical interactions with others seem to be carried into the virtual environment.[2] More profoundly, their studies indicate that bodily-based behaviors occurring in virtual exchanges with avatars of other people are occasionally *carried back* into subsequent real-world behaviors. For example, in online dating games, making an avatar more attractive than the user actually is not only boosted the user's self-confidence in social exchanges in the virtual world, but the improved self-perception also persisted outside of the virtual dating game in subsequent real-world social exchanges. Such participants, they conclude, "actually thought they had a shot with better-looking people" (Blascovich et al., 2011, 107) outside the virtual environment.[3] In sum, their research shows that people do not regard their avatars—nor do they regard the avatars of others—as mere representations or empty animations. Rather people inhabit their avatars much like they inhabit their own body, and this occurs not after prolonged use but within minutes of controlling an avatar.

Thus, anecdotally and experimentally, there appears to be a strong identification with one's avatar "body" as well as the transposition of constraints and possibilities of real-world embodiment and sociality into the virtual environment. I shall call this cluster of phenomena, "virtual re-embodiment." How are we to understand such experiences from a post-phenomenological perspective? In order to begin answering this question, I shall examine analogous experiences that occur to operators of tele-robotic systems, for there are crucial similarities between the two.

In an article several years ago, Jonathan Cole, Oliver Sacks, and Ian Waterman describe their experiences using a tele-robotic system at Johnson Space Center in Houston, Texas. Here is what they report:

> One sees and controls the robot's moving arms without receiving any peripheral feedback from them (but having one's own peripheral proprioceptive feedback from one's unseen arms). In this situation, we transferred tools from one hand to another, picked up an egg, and tied knots. After a few minutes we all became at ease with the feeling of being "in" the robot. Making a movement and seeing it effected successfully led to a strong sense of embodiment within the robot arms and body. This was manifest in one particular example when one of us thought that he had better be careful for if he dropped the wrench it would land on his leg! Only the robot arms had been seen and moved, but *the perception was that one's body was in the robot*. (Cole et al., 2000, 167; italics added)

I will call this sense of being "in" or "inhabiting" the remote body of the robot, "robotic re-embodiment." It occurs to varying degrees in tele-robotic systems, such as remotely operated vehicles and tele-surgery, for example. It seems that the effect is achieved as soon as the operator has real-time visual access to and feedback from the remote site and is able to affect change there, usually using some kind of robotic "arm" and/or grasping device. Approaching robotic re-embodiment from a post-phenomenological framework, it is tempting to see the remote robotic arms as technological extensions of the user's carnal embodiment, much in the same way that ordinary co-located tools extend our sense of embodiment as well as our bodily capacities for perception and action. While there are important differences, three similarities between virtual and robotic re-embodiment are critical. Both involve some kind of interface equipment (joystick, head-mounted display, keyboard, etc.), visual access to and feedback from the remote environment, and the ability to be active and effective there. Since both virtual and robotic re-embodiment involve expanded perceptual access and greater agency in the world, there appear to be good reasons for seeing virtual and robotic re-embodiment as technological extensions of carnal embodiment, much like the blind man's cane in Merleau-Ponty's famous description.

## TECHNOLOGICAL EXTENSIONS AND INCORPORATIONS

The *locus classicus* for beginning a post-phenomenological analysis of technologies as extensions of carnal embodiment is Merleau-Ponty's famous example of the blind man's cane in the *Phenomenology of Perception.* For the blind man, the cane is initially experienced as a perceived object, but after some practice, it withdraws from focal awareness, becoming a perceiving object—an almost transparent means *through which* the blind man senses the environment around him. Once mastered, the stick becomes an element in the blind man's motor-perceptual repertoire. In his description, Merleau-Ponty modulates between language implying that the cane becomes a bodily "extension" and language suggesting that the cane should be seen as a bodily "incorporation."[4] This ambiguity can lead to some confusion.

In order to avoid misunderstanding Merleau-Ponty's intention here, it is important to clarify the target of his descriptions. Taken most literally, the terms, "incorporation" and "extension," imply bringing something *into* the body or extending the body *outside* its physical boundary. This way of understanding Merleau-Ponty's language in these passages refers to the "objective body," that is, the body as a material object persisting in objective space and having an "inner" and an "outer" that is separated by a boundary. This boundary is sometimes provocatively referred to as the "skin-bag."[5] The examples Merleau-Ponty deploys in these passages and the language he uses to describe them support, to some degree, this reading.[6] But this cannot be a complete account of Merleau-Ponty's intention, for at issue here is not simply the objective body, but the "phenomenal body," that is, the lived body—the body as the locus of intentional activity in the world. More precisely, what is at stake in these passages from the *Phenomenology of Perception* is the relationship of habit and the "body schema" (*schema corporel*).

While I will have more to say about this below, it is crucial to note that while it is indeed supported by neurological and other physical structures, the body schema cannot be reduced to such structures.[7] From a phenomenological point of view, they are distinct; thus, descriptions of the "body schema" refer to a different level of embodiment in which different principles apply. Crucially, the organizational logic of the body schema differs from that of the objective body in one important way. Understood in terms of body schema, "incorporation" and "extension" do not imply the breaching of a boundary that would distinguish that which is "internal" to the body from that which is "external." Instead, what is primary here is a relationship between part and whole. This relationship is *not* one in which already existing discrete parts are externally and contingently related to each other by being brought into a whole through some additive principle[8]—like so many apples in a basket or so many interchangeable parts on a car—rather, the parts are related to each other and to the whole such that the organizing "logic" of the whole is a prior principle of organization that required just these parts in order to realize itself. In other words, referring to the body schema, the relationship between the whole body and its "parts" is akin to the relationship between the perceptual Gestalt and its various temporal "moments," and this relationship gives primacy to the whole.[9]

Understood as an alteration that requires a re-synthesizing of the entire body schema and its inter-related elements, once a technological artifact is mastered enough to withdraw from focal attention, it becomes integrated into the body schema such that a new body synthesis emerges. Incorporating a technological artifact into the body schema implies the emergence of a renewed body with expanded perceptual powers and extended capacities for agency in the world. What Merleau-Ponty's analysis of the cane shows is that insofar as the body is the locus of intentional activity in the world, it portrays an open structure that can be modified by integrating into itself potentialities and possibilities afforded to it by non-bodily objects—what I shall call, "tools"—in its environment. The incorporation of a tool in this manner implies that it not only withdraws from focal awareness but that the affordances provided by that tool become a background condition for subsequent intentional activity. In other words, the language of "incorporation" implies that the tools contribute to the structuring of the very environment in which they will be enlisted to bring intentional activity to fulfillment. By doing so, a tool extends one's capacities beyond what they would be in the absence of it.[10] It is in this manner that the language of "incorporation" and "extension" are used somewhat interchangeably in phenomenological and postphenomenological analyses of technology and embodiment, even though the terms are not synonymous. Thus, the ambiguity arising from Merleau-Ponty's language can be alleviated *somewhat* with textual analysis as long as we keep in mind that what is at stake is not tied exclusively to an inner-outer distinction that would apply to the object body but rather what is central is a distinct form of the part-whole relationship that applies to the phenomenological body, specifically a pre-personal synthesis of the body schema.

This brief textual reminder, however, is not sufficient to fully eliminate confusions that have emerged from Merleau-Ponty's description of the blind man's cane. Although it was never intended to be a description of human-technology relations more generally, it is often a starting point for philosophical analyses that seek to understand the altered sense of embodiment that occurs in habituated human-technology engagements. In such analyses, it is not unusual for technological "incorporation" to be used interchangeably with technological "extension"; however, technological developments since Merleau-Ponty's time—such as medical implants of various sorts, advanced prosthetics, and technologies of telepresence—call for a more refined analysis in order to avoid obfuscation. In short, not all technological extensions are incorporations. In a recent article, Helena De Preester has sought to clarify this difference by arguing for a more rigorous distinction between the incorporation of non-bodily objects into the body, which she understands as "prostheses," and objects that must be regarded as "mere" bodily extensions (2011). Further dividing these two categories (prostheses and extensions) into three subdivisions—perceptual, limb, and cognitive—she proposes conditions for the possibility of each. Sensitive to both cognitive science and phenomenological approaches to technological embodiment, her argument is cogent and compelling, and I largely agree with her conclusions, specifically her claim that "incorporation of non-bodily items into the body… is a difficult process and limited by a number of conditions of possibility that are absent in the case of "mere" bodily extensions" (De Preester, 2011, 135). Indeed, I have argued elsewhere that genuine bodily incorporations are rare.[11] For De Preester, both prostheses and extensions result in a transformed sense of embodiment, but incorporations must meet additional requirements, namely they must give rise to "a feeling of body ownership."[12] Of course, "mere" technological extensions also result in a transformed sense of embodiment insofar as they extend one's bodily capacities. It is the transformed sense of embodiment that *does not* amount to an incorporation that I want to focus on in the remainder of this chapter, for I believe more fine-grained distinctions are warranted here as well. More precisely, I will argue for a distinction between two different varieties of technologies as extensions of embodiment, those that can be integrated into the body schema and those that remain at the level of the body image. This distinction is especially important when we consider technologies of telepresence, such as tele-robotic systems.

## ROBOTIC RE-EMBODIMENT AND THE DEAFFERENTED SUBJECT

As noted above, robotic re-embodiment describes an experience that often occurs to tele-robotic operators in which they have the sense that they are "in" or that they "inhabit" the remote machinery. What is the nature of this technological re-embodiment? Without a doubt, much like the blind man's cane, tele-operators must first learn to use the bodily co-located interface equipment that controls the remote machinery so that the interface equipment withdraws from focal awareness. It is important to notice, however, that there is a double-technological embodiment occurring in robotic re-embodiment. In addition to integrating the interface equipment into one's body schema, operators also experience the remote "arms" of the robot as functional bodily elements. The experience of robotic re-embodiment rests on the fact that the remote machinery must also withdraw to some degree in order to become an almost transparent medium through which one's perception and agency "flows" into the remote environment.

I argue that this second withdrawal is *distantly different from* the way in which bodily co-located tools and equipment—such as the blind man's cane—recede from focal awareness to become integrated into the body schema. There is a decisive difference here, for bodily co-located tools become integrated into the body schema by offering robust tactile feedback *and thereby participating in somatic proprioception.* This is often *not* the case with remote robotic machinery. The remote "arms" of the robot do not participate in the bodily information systems that give rise to somatic proprioceptive awareness. In brief, proprioception refers to the awareness one has of one's own body in space. While the future promises remote machinery that will yield robust haptic feedback to the operator, currently this is not the case for many tele-operated systems. But, as contemporary technology shows, such feedback is not even necessary to produce a sense of remote telepresence and robotic re-embodiment, and this is where things get very interesting. For many tele-operated systems hitherto developed and currently in use produce the sense of robotic re-embodiment with only visual feedback from the remote environment and real-time command and control of the remote machinery. In other words, they produce a sense of technological re-embodiment without offering proprioceptive feedback for the operator. What does this show us?

To answer this question, it will be useful to look at the case of IW, a subject who has lost cutaneous touch and somatic proprioceptive awareness. IW's case has been well documented.[13] As a result of an illness at nineteen, IW suffered acute neuropathy that left him without any sense of touch or proprioceptive awareness below his neck. Sometimes referred to as an inner sense of one's own body, somatic proprioception describes the persistent pre-reflective awareness one has of one's own body position in space. It is an immediate, non-perspectival awareness of one's body. For example, due to proprioception, one can point to one's ankle without having to look down to find it; this is so, even in the dark. Moreover, due to somatic proprioception, I have a continual pre-reflective awareness of my body throughout its movements; such awareness is essential for motor control. While it depends to some degree on input from vision and the vestibular system, somatic proprioception relies primarily on information from "kinetic, muscular, and cutaneous sources" (Gallagher, 2005, 45). It was precisely these sources that were destroyed for IW. While he still can experience hot, cold, pain, and muscle fatigue in areas below the neck, he has lost any sense of cutaneous touch and has "no proprioceptive sense of posture or limb location" (Gallagher, 2005, 43).

This malady affected him greatly. At the onset of his illness, he lost the ability to stand, walk, or even achieve coordinated motor control.[14] After the injury, it took him several months to regain the ability even to stand. Eventually, he relearned to walk, and after a few years, was able to regain enough motor control to complete daily tasks and slowly resume living a somewhat normal life. Doing so, however, requires an altered form of human embodiment, for without proprioceptive information continuously monitoring and updating his body's position in space, coordinated bodily movement for IW requires constant visual attention and mental concentration. For example, walking across uneven ground is very challenging for IW, requiring most of his attentive awareness. By contrast, in normal, non-pathological bodily engagement with the world, the body disappears. The body effaces itself in normal activity. While going about one's daily tasks, the body, while always marginally distinguishable from its environment, becomes woven into the texture of the world. Reaching for a coffee cup, for example, does not require a conscious command that one's arm then carries out with continuous visual focus remaining on the location of the arm in space throughout its movement. Rather, most of the specific motor activity of arm movement occurs preconsciously due to almost automatic motor-perceptual processes that have been habitually sedimented in the body. Focally, one is intentionally oriented towards the coffee cup *not* the arm that achieves the movement. But without proprioception, the automatic responses allowing the body to efface itself in intentional activity have been disrupted for IW. Thus, in any movement the he initiates, he must focus on his body in an unusually attentive manner. Gallagher and Cole describe IW's mode of embodiment:

> Without proprioceptive and tactile information [IW] knows neither where his limbs are nor controls his posture unless he looks and thinks about his body. Maintaining posture, is for him, an activity rather than an automatic process.

> His movement requires constant visual and mental concentration. (Gallagher and Cole, 1998, 135)

For IW, successful performance of daily activities such as walking, sitting upright, and eating require much greater reliance on active perception of and attentive concentration on the body. Additionally, "motor schemas"—those habitually constituted automatic responses to a stereotypical situation, such as buttoning a shirt—are largely absent for IW. Moreover, he even seems incapable of generating new ones. Thus, IW's daily getting about in the world relies much more on active attention (primarily visual) and less on sedimented bodily habits. So, what does IW's case reveal about robotic re-embodiment?

I believe IW's case highlights four things that are relevant to understanding robotic re-embodiment. First, following Gallagher's assessment of IW's case, it shows us that there is a clear conceptual distinction between body image and body schema as well as a functional difference. From a phenomenological point of view, the human body is located on both sides of the intentional relationship. The difference between "body image" and "body schema" is meant to articulate this difference. Body image consists of "a complex set of intentional states and dispositions—perceptions, beliefs, and attitudes—in which the intentional object is one's own body" (Gallagher, 2005, 25). Thus, the body image refers to bodily self-awareness in a self-reflexive form of intentionality, and it involves perceptual, conceptual, and affective modes. For example, looking at oneself in the mirror, dispassionately thinking about one's body in a medical context, or finding one's body disgusting because it does not meet social norms of beauty are all instances of self-reflexive intentional states of the body image. In all such cases, one's body is taken as an object of an intentional regard. In contrast, "body schema" describes a complex system of "sensory-motor functions that operate below the level of self-referential intentionality… preconscious, subpersonal processes that play a dynamic role in governing posture and movement" (Gallagher, 2005, 26). "Body schema" refers to a pre-reflexive synthesis of inputs from multiple bodily systems—proprioceptive, visual, and vestibular—that structures, constrains, and enables intentional bodily activity (Gallagher, 2005, 132–152). Thus, although conceptually distinct, body schema and body image are often functionally interrelated. The body schema undoubtedly supports normal, everyday intentional activity. In fact, it is the accomplishment of the sub-personal body processes that allows one to complete habitual movements almost automatically, thereby allowing the body to efface itself in normal intentional activity. Likewise, attentive awareness of one's body position (body image) can contribute to alterations in the body schema; for example, when one is learning a new dance move, focusing on the body's position and movement is necessary for the move to become habitual and sufficiently sedimented in the habit body so that one is able to execute the move almost automatically. Of course, this functional integration of body image and body schema is operative in learning to use bodily co-located tools. The notion of *withdraw* from focal awareness that occurs as one becomes habituated to the motions and movements of a hammer, for example, expresses the shift to a primarily pre-reflective (body schema) mode of engagement.

The second important consideration that IW's case highlights is that even though inputs from both the vestibular system and visual sense contribute to the synthesis of a body schema, proprioceptive information from kinesthetic, muscular, and cutaneous sources is *essential* to form the body synthesis that gives rise to a sense of bodily unity. In fact, insofar as proprioception is a pre-personal bodily self-awareness, we can speak of "proprioceptive spatiality of the body" (Gallagher, 2006, 351). Distinct from allocentric space—sometimes referred to as "objective" space—in which spatial ordering is keyed to an object outside the body, proprioceptive spatiality is an intra-corporeal unity in which ordinary spatial relations (based on a notion of extension between objects) do not apply. Merleau-Ponty puts it this way: "bodily space can be distinguished from external space and it can *envelop* its parts rather than laying them out side by side" (Merleau-Ponty, 2012, 103, italics added).

I noted above that Merleau-Ponty's analysis of body schema implies a particular part-whole relationship that resists the inner/outer distinction; it also resists notions of "near" and "far" understood as ways of describing spatial distance in which discrete things are separated by extension. It makes no sense to say that my nose is nearer to me than my right big toe. When it comes to proprioceptive spatiality of the body, there is no center; there is no origin. There is the body as a diversity-in-unity. Such a system of organization is a differentiated unity of mutual self-envelopment and overlap by the various elements or parts. Given this notion of spatiality, the diversity-in-unity of the body schema *abolishes distance* not merely in the sense that my body is always "here" in its entirety—even as it co-opts tools—but, more importantly, proprioceptive spatiality abolishes distance also in the sense that the very notion of distance cannot be applied to the body thusly regarded. So, to speak of bodily co-located tools becoming an extension of embodiment implies an absence of distance—not merely a factual absence but rather that the very concept of distance *does not apply* to the resulting body-tool unity. The same cannot be said, however, of tele-operated robotic systems, for not only is there an objective distance between the operator's physical body and the remote elements of the system, but also the conceptual dyad, near/far (understood in terms of extension between objects) characterizes a salient aspect of the primary mode of engagement with these technologies.

The third thing that IW's case highlights is that even though proprioceptive information is an essential aspect of body schema syntheses, deficiencies in proprioceptive receptors can be compensated for, to some minimal degree, with attentive vision and active concentration. In other words, with great effort, one can cope with loss of normal functioning of somatic proprioception and cutaneous sensation, but doing so alters the subject's embodied engagement with his or her environment in a crucial way. As Gallagher and Cole describe it:

> In place of missing body schema processes, IW has substituted cognitively driven processes that function only within the framework of a body image that is consciously and continually maintained. If he is denied access to a visual awareness of his body position in the perceptual field, or denied the ability to think about his body, then is motor control ceases to function. (Gallagher et al., 1998, 140)

For IW, location of the limbs in space and intentional bodily activity are no longer elements of his pre-conscious, pre-personal body schema but become elements of his conscious, self-referential body image. Even routine movements will likely never become semi-automatic process that can recede to the margin of conscious awareness but will always require focal concentration and active visual attention.

The fourth thing that IW's case highlights—and crucial to my overall argument here—is that the state of the tele-operator is an *analogous* state as IW when it comes to the remote machinery. This is so because the tele-operator only has visual sensations—and not tactile, kinesthetic, or proprioceptive awareness—of the remote machinery's location in and movement through space. Thus, activity in the remote environment would involve a mode of embodiment similar to IW's experience of his own body below his neck. This means that in any remote activity undertaken with the tele-operated system, the operator must attentively focus on the target object of the intended action as well as the remote machinery that serves as the means with which the intended action is brought to fulfillment. In tele-surgery, for example, the surgeon must focus on the "tissue" of the patient as well as on the position and movement of the robotic arms. Like IW, activity in the remote environment requires constant visual and mental concentration.

Even with this limit, however, visual feedback can yield what has been called "visual proprioception" (Gallagher and Cole, 1998, 143), which is the sense that one has of the nearness and farness of objects in one's visual field as one moves about the environment. One must do so with one's object body as the constant orienting referent. The most obvious instance of visual proprioception involves driving a car. For an experienced driver, successfully guiding a car to the desired destination can be achieved with minimal explicit attentive awareness of the spatial boundaries of the car as well as the precise location of nearby objects. Indeed, it is occasionally the case that when driving a routine route, such as going home from work, one "zones out," being pre-occupied with the day's business or upcoming nighttime activities; nonetheless, one keeps the road ahead and objects to be avoided at the margins of one's attentive visual focus. Although vision is powerful enough to compensate somewhat for the absence of proprioceptive feedback, we must be careful not to understand any withdraw of the remote machinery of telerobotic systems to be full integration into the operator's pre-conscious body schema on the same order as the withdraw of the bodily co-located interface equipment. Rather, when it comes to the remote machinery, the sense of re-embodiment experienced by tele-operators involves the bodily self-awareness indicative of the body image—as the object of an intentional regard. Much like IW's bodily situation post-neuropathy, we can even expect that over time few, if any, automatic motor actions with the robotic "arms" would emerge for the operator, while such motor schemas would likely emerge with use of the bodily co-located interface equipment.

To summarize the point of my argument in this section: without robust tactile feedback and the ability to participate in proprioceptive information sources, there is a *decisive dis-analogy* between the experience of robotic re-embodiment and experiences with bodily co-located tools such as the blind-man's cane. In short, the manner in which the tele-operator embodies bodily co-located interface equipment is *different in kind* from the manner in which she re-embodies the remote robotic "arms." Thus, the example of the blind-man's cane, understood as a technological extension of carnal embodiment, reaches its limit when making sense of robotic re-embodiment. Moreover, since the possibilities for agency and perception offered by the remote elements of the system cannot be integrated with the pre-personal body schema but must remain at the level of body image, they should not be regarded as extensions of embodiment in the same manner but are technological extensions of embodiment in a fundamentally different sense. This is even more evident when we consider virtual re-embodiment.

## VIRTUAL RE-EMBODIMENT

Having taken a detour through robotic re-embodiment, I am now prepared to return to virtual re-embodiment. In robotic re-embodiment, visual access to and feedback from the remote location along with the ability to manipulate the remote elements of the robotic system give the operator a sense of presence in the remote environment. It shares these three aspects with virtual re-embodiment: some kind of interface equipment (joystick, head-mounted display, keyboard, etc.), visual access to and feedback from the remote environment, and the ability to be active and effective there. I would like to suggest that there is another, deeper similarity between the two, for in both we witness the same *double technological embodiment*. The tele-operator, like the game player (or participant in an immersive virtual environment) must become familiar enough with the bodily co-located interface equipment so that it withdraws from focal awareness, becoming integrated into the pre-personal body schema. Moreover, it often occurs that the tele-operator experiences the sense of inhabiting the remote machinery, despite the lack of tactile feedback from the remote environment.

As noted above, this sense comes about due to visual feedback as well as real-time command and control of the remote robotic "arms." Likewise, we can conclude that *in an attenuated but analogous fashion* one's avatar, which is seen and manipulated but never tactilely sensed, gives rise to the sensation that one inhabits the avatar "body" in the virtual environment. Furthermore, it seems that manipulating an avatar in a virtual environment requires the same attentive visual focus as the tele-operation of remote machinery. This means, however, that the manner in which one embodies the co-located interface equipment in virtual embodiment *is different in kind* from the experience one has of inhabiting the avatar "body," for the interface equipment—joystick, VR helmet, or whatever is used—can withdraw into one's body schema to become an almost transparent medium of one's intentional activity, while the same cannot be said of the digital representation that is one's avatar. The avatar remains at the object end of the (visual) intentional relationship, even though, much like the remote machinery of a tele-robotic system, it functions as the locus of perception and agency in the virtual environment.

The similarity to robotic re-embodiment elucidates a crucial aspect of virtual re-embodiment. Moreover, it also hints at the *limitations* of that re-embodiment, for if the remote elements of the robotic system should not be seen as extension of carnal embodiment—as long as this is understood in reference to something like the blind-man's cane—then even less so can one's avatar be seen as an extension of embodiment. Rather, much like robotic re-embodiment, it is much closer to a pathological form of embodiment evinced by IW. In other words, while one might identify with one's avatar, one does so in a self-referential manner indicative of an intentional body image. In no sense can it be understood that one's avatar comes to be integrated into one's pre-personal body schema but, then, neither do the remote "arms" of the tele-robotic system. The similarities between these two varieties of technological re-embodiment are quite deep.

Of course, there is one glaring difference between virtual and robotic re-embodiment, namely that in the latter, the remote environment is a location in the actual, physical world, even if it happens to be on Mars. Drone pilots drop real bombs and tele-surgeons save real lives. Contrariwise, the remote environment in virtual embodiment is an imaginary one; perception and action occur in fantasy time and fantasy space. While virtual re-embodiment entails a strong imaginary element, the reality principle remains, at least in affective terms, for playing video games or otherwise spending one's time in virtual environments leads to *real* emotional experiences. Real people fall in love with other real people in Second Life, for example. Making sense of such emotional experiences, however, is a topic for another chapter.

## CONCLUSION

In this chapter, I pursued a postphenomenological analysis of two varieties of technological re-embodiment, robotic and virtual. Technologies of telepresence such as tele-robotic systems, advanced video games, and immersive virtual environments offer real-time visual access to and agency in remote environments—whether they are real or virtual—thereby providing users of these technologies with a sense of being an active, perceiving agent both "here" and "there." While they may expand our perceptual access to the world and give us greater agency, I have argued that we should avoid thinking of remote robotic arms and avatar "bodies" as *technological extensions of embodiment* on the same order as bodily co-located tools such as the blind man's cane or the carpenter's hammer. Doing so implies an abolition of distance and integration into the body schema that does not, in fact, occur. Given the fact that the player/operator's mode of engagement with the "remote" elements of these technologies is primarily visual, without participating in pre-personal proprioceptive bodily systems, the level of bodily integration seems limited to that of the body image. In short, the "remote" elements remain at the noematic terminus of the intentional relationship. Therefore, a distinction can and should be made between these two types of technological extensions: those that are assimilated into the body schema and those that are integrated into the body image. This distinction provides a needed refinement of the extension thesis, which is the *general* claim that technologies are extensions of embodied human perception, agency, and cognition. The ability (or inability) of a technology to contribute to proprioceptive self-awareness is crucial to the distinction I am making.

Future technologies might promise the theoretical possibility of interface equipment that can participate in proprioceptive information systems of the body; however, it seems that multiple practical possibilities would need to be surmounted first. Indeed, such promises seem to be more on the order of techno-fantasies rather than plausible developments of current technological trajectories, which seem to point in the directions of greater visual resolution, more "natural" interface apparatuses, shorter feedback times, and greater mobility and dexterity in remote environments. Even without radical advances, however, current technologies of telepresence offer sufficient visual representation along with real-time command and control in and feedback from a remote environment to induce a sense of being an active, perceiving agent both "here" and "there."

Perhaps this doubling of perception and agency offers us a better way to understand the re-embodiment experienced with technologies of telepresence, namely, as an asymmetrical doubling of embodiment, with one's carnal embodiment remaining an anchor, and, thus, any sense of agency and perception in the remote environment will be attenuated and will retain a persistent reference back to carnal embodiment as the locus of intentional activity in the world. Understanding such experiences as an asymmetrical doubling can account for the continual attentive shifting between agency and perception in the local and remote environments that occurs with the use of technologies of telepresence. After all, one cannot "shift out" of one's carnal body. It is a permanent anchor of one's embodied situation.

## NOTES

1. See, for example (De Preester, 2011; Dolezal, 2009; Ihde, 2011). "Re-embodiment" has come to take on two related but distinct meanings. On the one hand, "re-embodiment" might be applied to the altered form of embodiment emerging from the integration of bodily co-located "tools" into the body schema, such as we witness in the example of the blind man's cane or the carpenter's hammer. On the other hand, "re-embodiment" also can refer to the altered sense of embodiment that occurs with technologies of telepresence in which technological equipment yields a sense of perception and agency in a remote environment, whether that environment is robotic or virtual. I will use this term in the second sense.

2. The subjects of these tests wore head-mounted displays so that they could see nothing of the actual physical room they were in, and they were moving about a physical room that corresponded to the virtual room in which their avatar was located. (See Bailenson et al., 2003; Blascovich and Bailenson, 2011, 86–89.)

3. See also (Bailenson and Yee, 2009). In a series of studies, Bailenson and his colleagues examine what they call the "Proteus Effect," which is the thesis that an individual's behavior in virtual environments conforms to their digital self-representation independently of how others see their avatar. While it seems reasonable that people will behave more aggressively in virtual exchanges with others while controlling an aggressive-looking avatar, what is surprising is that some of these altered behavioral traits persist, at least for a short time, into subsequent real-world social engagements.

4. For example: "To habituate oneself to a hat, an automobile, or a cane is to take up residence in them, or inversely, to make them participate with the volumnosity of one's own body. Habit expresses the power we have of dilating our being in the world, or of altering our existence through incorporating new instruments [Fr: *annexant de nouveau instruments,* 168]" (Merleau-Ponty, 2012, 145). Also, for the blind man, the cane "is an appendage of the body, or an extension of the bodily synthesis [Fr: *C'est un appendice du corps, une extension de la synthèse corporelle*, 178]" (Merleau-Ponty, 2012, 154).

5. This way of describing the physical body appears in "extended mind" approaches to embodiment in cognitive science. See for example, Andy Clark's 2003 book, *Natural Born Cyborgs*, for examples—and it is often used in the phrase "biological skin-bag" (pp. 16, 33, and 44).

6. Other examples he uses include a ladies hat with a large external feather, a car, a typewriter, and a musical organ.

7. The distinction between "body schema" and "body image" and the role proprioception plays in framing this difference are central to my argument in this paper, and I will describe this in more detail below.

8. "The contour of my body is a border that ordinary spatial relations do not cross. This is because the body's parts relate to each other in a peculiar way: they are not laid out side by side, but rather envelop each other" (Merleau-Ponty, 2012, 100).

9. For a further discussion of gestalt-like unities, see my *Merleau-Ponty's Phenomenology: The Problem of Ideal Objects* (Besmer, 2008, 21–27).

10. It so happens that the blind man's cane also extends his physical body; however, it is the enlargement or expansion of possibilities for intentional action and fulfillment that is central to the notion of extension here.

11. I pursue this question by examining the concept of "cyborg" and "cyborg intentionality" (Besmer, 2102).

12. The full quotation is: "Thus, my exact claim is that the distinction between bodily extension and body incorporation is based on a feeling of body ownership" (De Preester, 2012, 396).

13. IW's case is detailed in Jonathan Cole's book, *Pride and a Daily Marathon* (1995). It is also documented in Shaun Gallagher's 2005 book, *How the Body Shapes the Mind*. IW's case is also described in an essay co-authored by Cole and Gallagher (Gallagher and Cole, 1995). This article is reprinted in (Gallagher and Cole, 1998). My references will be to the reprinted edition. For a discussion of a similar case (GL) see also (Cole and Paillard, 1998). Oliver Sacks describes a similar instance in (1985, 42–52).

14. Hospitalized immediately after the onset of the neural damage, IW's first sense was one of *utter disembodiment*. Cole describes his condition: "[IW] seemed to be "floating" on the mattress. Without sense of position or touch from his body and limbs, he appeared not to be resting on the bed. But it wasn't the relaxed floating one associates with swimming… but an almost unimaginable total absence of feeling" (Cole, 1995, 14).

## REFERENCES

Bailenson J., and J. Blascovich. (2011). *Infinite Reality: Avatars, Eternal Life, New Worlds, and the Dawn of the Virtual Revolution*. HarperCollins: New York.

Bailenson, J., J. Blascovich, A. Beall, and J. Loomis. (2003). "Interpersonal Distance in Immersive Virtual Environments." *Personality and Social Psychology Bulletin* 29 (7): 819–833.

Bailenson, J. and N. Yee. (2009). "The Difference Between Being and Seeing: The Relative Contribution of Self-Perception and Priming to Behavioral Changes via Digital Self-Representation." *Media Psychology* 12 (2): 195–209.

Besmer, K. (2008). *Merleau-Ponty's Phenomenology: The Problem of Ideal Objects*. Continuum: New York.

Besmer, K. (2012). "Embodying a Translation Technology: The Cochlear Implant and Cyborg Intentionality." *Techné: Research in Philosophy and Technology* 16 (3): 293-316.

Blascovich J., and J. Bailenson. (2011). *Infinite Reality: Avatars, Eternal Life, New Worlds, and the Dawn of the Virtual Revolution*. New York: HarperCollins.

Clark, A. (2003). *Natural Born Cyborgs*. Oxford UP: New York.

Cole, J. (1995). *Pride and a Daily Marathon*. Cambridge, MA: MIT Press.

Cole J. and J. Paillard. (1998). "Living without Touch and Peripheral Information about Body Position and Movement: Studies with Deafferented Subjects." In J. Bermúdez (ed.), *The Body and the Self*. Cambridge, MA: MIT Press, 245–266.

Cole J., O. Sacks, and I. Waterman. (2000). "On the Immunity Principle: A View from a Robot." *Trends in Cognitive Science* 4(5): 167–68.

De Preester, H. (2012). "Technology and the Myth of 'Natural Man.'" *Foundations of Science* 17 (4): 385–390.

De Preester, H. (2011). "Technology and the Body: the (Im)Possibilities of Re-Embodiment." *Foundations of Science* 16 (2): 119–137.

Dolezal, L. (2009). "The Remote Body: The Phenomenology of Telepresence and Re-Embodiment." *Human Technology: An Interdisciplinary Journal on Humanities in ICT Environments* 5 (2): 208–226.

Gallagher, S. (2005). *How the Body Shapes the Mind.* Oxford UP: New York.

Gallagher, S. (2006). "The Intrinsic Spatial Frame of Reference." In Dreyfus, H. and Wrathall, M. (eds.), *Blackwell Companion to Phenomenology and Existentialism.* Malden, MA: Blackwell, 346–355.

Gallagher, S., and J. Cole. (1995). "Body Image and Body Schema in a Deaffernted Subject." *The Journal of Mind and Behavior* 16 (4): 369–90.

Gallagher, S., and J. Cole. (1998). "Body Image and Body Schema in a Deafferented Subject." In D. Welton (ed.), *Body and Flesh: A Philosophical Reader.* Oxford: Blackwell, 131–147.

Ihde, D. (2011). "Postphenomenological Re-Embodiment." *Foundations of Science* 17 (4): 373–377.

Merleau-Ponty, M. (1945). *Phénoménology de la Perception.* Librairie Gallimard: Paris.

Merleau-Ponty, M. (2012). *Phenomenology of Perception.* trans. Donald A. Landes. New York: Routledge.

Sacks, O. (1985). "The Disembodied Lady." In *The Man Who Mistook His Wife for a Hat.* New York: Touchstone, 42–52.

## *Chapter Four*

# Thinking Technology with Merleau-Ponty

Aud Sissel Hoel and Annamaria Carusi

In this chapter we seek to expand the conceptual toolbox of postphenomenology by pointing to untapped resources for understanding human-technology relations in the later thinking of Maurice Merleau-Ponty. Merleau-Ponty is frequently acknowledged for his profound and groundbreaking account of embodied perception and action in *Phenomenology of Perception* (1962),[1] which has become a touchstone in matters relating to the body and embodiment across disciplines (Sobchack, 2004; Cataldi and Hamrick, 2007; Buchanan, 2008). Merleau-Ponty has also become a key reference in current attempts to rethink knowledge and experience in dynamic terms (Barbaras, 2004; Carbone, 2004; Toadvine, 2009), including attempts to develop relational ontologies (Low, 2000; Bannon, 2011). He is less frequently acknowledged for being an innovative thinker about technologies. However, as we have argued elsewhere (Carusi and Hoel, 2014b; Hoel and Carusi, forthcoming), critics who point to Merleau-Ponty's seemingly scant treatment of technologies (Latour, 1999, 9–10; 2003, 16–17; Ihde and Selinger, 2004, 361–367) tend to overlook the extent to which instruments, tools, and technologies were a constant preoccupation of his. This is particularly evident in his later thinking, as he shifted decisively away from a classical phenomenology. Artificial symbolisms and artifacts take center stage in the "indirect ontology" developed in *The Visible and the Invisible* (1968).[2] The expressive capacities of instruments and technologies are further explored in unfinished manuscripts and lecture notes (Merleau-Ponty, 1973; 2003; 2011).

In this chapter we aim to draw attention to the unrealized potential of the oeuvre of Merleau-Ponty to give a novel account of technological mediation. The later thinking of Merleau-Ponty is characterized by the way that the investigation of the perceiving body converges on an ontological exploration that acknowledges the ontological import and transformative capacities of a broad array of mediating apparatuses (the bodily apparatus, art works, language and other symbolic systems, tools, algorithms). In the following, we hope to demonstrate the relevance of Merleau-Ponty's indirect ontology to some of the key concerns of present-day postphenomenology, and the extent to which an engagement with Merleau-Ponty's expansive and dynamic notion of "flesh" may serve significantly to deepen our understanding of our interactions with technologies, including computation. Like today's postphenomenologists, the later Merleau-Ponty is concerned to show both that the body is technologized and that technologies are embodied—hence, the continued relevance of phenomenological frameworks. Some further important points of intersection include:

(1) The ambition to reconceive technological mediation in non-instrumental and non-representational terms: Postphenomenology is concerned to show that human activity is, and has always been, technologically embodied (Ihde, 1990, 20), and therefore technologies cannot be considered external instruments or representations standing proxy for perception and cognition.

(2) The focus on the transformative and non-neutral roles of technologies in knowledge and experience: The mediation by technologies does not occur between preformed entities, but instead plays a role in the co-constitution of both sides of the subject-object relationship. Technologies are seen as transformative of both perception and phenomena (Ihde, 2002, Selinger, 2006; Verbeek, 2008).

(3) The development of new and dynamic conceptual frameworks for understanding human-technology relations: Postphenomenology seeks to conceptualize relationships between human beings and technologies, showing them to be closely interrelated without obscuring their ontological differences. Technological mediation always involves what Ihde refers to as and "amplification/reduction structure of the human-technology experience" (Ihde, 1990, 78).

(4) The call for a new kind of technoscience criticism: Whereas traditional philosophy of science often fails to examine critically the role of technologies in scientific contexts, the postphenomenological approach takes seriously the technological embodiment of science in its instrumentation and apparatus. It does so without romanticizing unaided human experience and handicraft technologies, and without reverting to technological determinism. While we claim that this last point is a point of convergence between Merleau-Ponty and postphenomenology, this goes against a common reading of

Merleau-Ponty as exemplifying the romanticizing trend (see for example Ihde and Selinger, 2004). However, we hold that a different reading of Merleau-Ponty's relationship to science is possible, a point to which we return in the concluding paragraph.

The perspective on Merleau-Ponty that we put forward is based upon our ongoing work drawing out the implications of Merleau-Ponty's later thinking for understanding scientific vision (Carusi and Hoel, 2014a; 2014b; Hoel and Carusi, forthcoming). It may be surprising to evoke Merleau-Ponty in such an account, especially since Merleau-Ponty at times seems more concerned to contrast a negative appraisal of scientific vision by positive appraisals of other forms of vision—especially the vision of painting—than to build an account of scientific vision as such. In the influential essay "Eye and Mind," for example, he seems to be critical of modern science and the technological artificiality of its methods, whose phenomena are "more likely produced by the apparatus than recorded by it" (Merleau-Ponty, 1993, 122). He is likewise scornful in his judgment of the ideology of cybernetics, "where human creations are derived from a natural information process, itself conceived on the model of human machines" (1993, 122). However, as we have noted elsewhere (Carusi and Hoel, 2014b), the target of Merleau-Ponty's criticism is not so much science *in practice* but a certain way of *thinking* about science, which he associates with a particular strand of Cartesian thought and refers to as "technicized" and "operational" thought (1993, 137). This is a thinking that relates to phenomena in an external way, or, in a phrase reminiscent of Donna Haraway's explication of the "god trick" (Haraway, 1988, 581), a thinking that "looks on from above" (Merleau-Ponty, 1993, 122).

While Merleau-Ponty did not dedicate whole articles or books to the topic of technology, his later works and lecture notes are peppered with references to instruments, tools, and technologies. Significant sources include the unfinished manuscript *The Prose of the World* (1973), which he commenced immediately after *Phenomenology of Perception*, with the aim of showing how the account of perception given there could be related to the intellectual and cultural world; and the 1953 lecture notes published in *Le Monde sensible et le monde de l'expression* (2011), which show a strong preoccupation with symbolic systems, on one hand, and instruments, tools, and technologies on the other. In these writings, it is clear that he considers tools and artifacts not just as the expressions of humans, but as themselves expressive of objects or the world (2011, 48, 54). He also clearly points to the transformative effect of tools and works of art, frequently referring to both kinds of artifacts in the very same sentences (for example, 2011, 48, 53, 54, etc.). In "Eye and Mind"—which is his most developed account of vision in art (and, we would argue, of science)—he again refers to "technical objects, such as tools and signs" in the same breath as he discusses the role of mirrors in painting

(1993, 129). Our point is that the continuity that Merleau-Ponty saw between the "natural" symbolism of perception, and "artificial" symbolisms and artifacts is the key to understanding his deeper account of technology. However, to the extent one can speak of a systematic approach to technology, this approach must be teased out of the scattered references to instruments, tools, and technologies in his later works, and in the notes and course lectures that surround these works and provide a broader context for them. Another crucial source in this respect are the written traces of Merleau-Ponty's lecture courses on the concepts of nature during the late 1950s (2003), where the later sketches contain an increasing number of references to apparatuses, instruments, and algorithms, as well as to the eye or vision as a computer.[3] Merleau-Ponty died at a young age, and it is unknown what systematic account he might have given of technologies. Even so, it is clear that he certainly did think about technologies, and was concerned to give an alternative account to the Cartesian-inspired account that still prevailed in the mid-1900s, and indeed still does in many quarters. In addition to this, it is clear that, with respect to the expressive capacities that the later Merleau-Ponty attributes to instruments and tools, questions of relevance to technology are to a large extent treated by him in the context of symbolic systems and art works. Thus, in our ongoing engagement with Merleau-Ponty's later work, we develop an account of technologies that is consistent with Merleau-Ponty's main lines of thinking (as much as we can grasp them through reading together his published and unpublished writings), while at the same time further developing some of his main ideas beyond what he actually wrote. We propose this as a possible Merleau-Pontian philosophy of technology, a "thinking with" Merleau-Ponty about technologies.

In the following we outline the main points of our account. Before we proceed, however, we want to accentuate an important insight to be gained from the way that Merleau-Ponty handles technologies in the context of other preoccupations such as body schema, movement, nature, art, language, and history. This shows the extent to which he saw technologies as interrelated with these other phenomena, or to use his own term, as "intertwined" with them. For example, in "Eye and Mind" he writes:

> Visible and mobile, my body is a thing among things; it is one of them. It is caught in the fabric of the world, and its cohesion is that of a thing. But because it moves itself and sees, it holds things in a circle around itself. Things are an annex or prolongation of itself; they are incrusted in its flesh, they are part of its full definition; the world is made of the very stuff of the body. (1993, 124–125)

The familiar examples, from the *Phenomenology of Perception*, of the stick of a blind man and the feathers on a woman's hat, demonstrate the way in which tools and other artifacts can become extensions of the body. However, whereas *Phenomenology of Perception* conceives tool-use in terms of *internalization* in the body schema; "Eye and Mind" conceives it in terms of a productive *displacement* of the body's capacities and its relationships with the world. In the latter case, the incrustation of things in the flesh, is understood to reconfigure the interrelationship between bodies and things. To trace the lines of thinking that leads to this position and to make sense of it, we start from Merleau-Ponty's critique of the Cartesian notions of vision and space and the particular form of representationalism associated with it.

The criticism of science put forth in "Eye and Mind" targets a widespread conception of science, sometimes implicit and sometimes formalized into a philosophy of science, which Merleau-Ponty refers to as "operationalism" (1993, 122–23, 137). In this conception, scientific vision is conceived in representational terms, "as an operation of thought that would set up before the mind a picture or representation of the world" (1993, 124). Operationalism thus conceived is a mistaken thinking about science, which Merleau-Ponty traces back to Descartes, whose theory of vision fails to recognize the internal complicity between vision and world. It fails, that is, to recognize the crossover and mutual interdependence between seeing and seen, knowing and known. By pursuing this Cartesian motif, science forgets the "primordial historicity" of the body and object; and due to this forgetfulness, it fails to account for how it gains access to its world.

Merleau-Ponty, by contrast, conceives of vision as "laboring" in a double sense of the word: it involves active work and it brings new things into being; vision and knowledge are generative processes. Merleau-Ponty's fascination with painting, and with modern painting like Cézanne's in particular, was motivated by the way painting shows the labor of vision, and hence, by his belief that painting is an important means for revealing the nature of vision. Thus, whereas painting expounds the true "poetic" nature of vision, science conceives vision as "prosaic" (1993, 142–144), that is, as secondary and external to the things whose identities are conceived as pre-given. The latter conception of vision he believes to be molded by a way of thinking about space, which again was epitomized by Descartes. For Descartes, as extended beings, we are in space, but we can only truly *know* space when we abstract our bodily being from it; we know space only through reason and hence as an ideal mathematical space. This externality of an embodied viewpoint to space is captured in the often misleading analogy between perspectival paintings and windows,[4] which implies that the two-dimensional surface of the painting opens onto depth as it would exist for bodiless beings. Thus understood, Merleau-Ponty observes, "the window opens only upon *partes extra partes*, upon height and breadth merely seen from another angle—upon the absolute positivity of Being" (1993, 134). It is important to note that Merleau-Ponty does not in fact criticize Descartes for having idealized space or for having proposed a notion of space that is a construct. On the contrary, he sees this stage of idealization and construction as a necessary step in the "freeing" of space, allowing for a different thinking about it than what appears to be accessible empirically. Rather, his criticism is targeted only at one point, namely, where this construction is conceived as a positivity and treated as though it were the only "real" space. Thus, what characterizes "operationalist" science is that it appropriates the mathematical construct as an exhaustive notion of all kinds of space, and assumes the complete certainty of its own contact with the world.

The core of Merleau-Ponty's criticism of "operationalism" in science is its assumption of externality between vision and things. This is an insuperable barrier to its acknowledgment of the complicity between them, and to seeing them as inextricably intertwined with each other—and as we argue, with symbolisms and technologies as well. Merleau-Ponty counters externality wherever he finds it (including Sartre's positing of the in-itself and for-itself). For Merleau-Ponty, it is necessary to see subjectivity, the body, and vision as operating "from the middle (*du milieu*) of things" (Merleau-Ponty 2008, 48, our translation); there is no privileged viewpoint "from above." In *The Visible and the Invisible* Merleau-Ponty acknowledges that the account he gave of perception in *Phenomenology of Perception* was still too reliant on the subject-object dichotomy;[5] "flesh" is the term that he uses for his new starting point. On Merleau-Ponty's account, flesh does not fit into established ontological categories and marks a break with dualist metaphysics. Flesh is neither substance nor consciousness; rather, it is "the formative medium of the object and the subject" (Merleau-Ponty, 1968, 147). Thus, flesh is "in the middle" also in the sense of serving as an operative, organizing force. The body is at once enmeshed in fleshly relations with objects, *and* itself an instantiation of flesh in the intertwining of phenomenal body and objective body (Merleau-Ponty, 1968, 179–180). Here it is important to note that the French term "milieu" comes into play in several meanings: "middle," "medium," and "environment." For, as Merleau-Ponty conceives it, each mediated situation opens a new dimension of the world, a specific "environment" with its own range of possible appearances and actions—or, in scientific contexts, a specific "space" of possible determinations and comparisons.[6] Merleau-Ponty's reconceptualization of the relationship between being and world in terms of flesh initiates a profound rethinking of mediation. Flesh does not mediate between pre-existent and independently constituted entities; rather, it is in and through mediation that entities are interrelated and become the entities they are. The mediation of flesh, in other words, is productive and formative; it has ontological consequences (hence the convergence in the later thinking of Merleau-Ponty of theory of vision and ontology).

With the notion of flesh as productive and formative mediation we arrive at the shift between *Phenomenology of Perception* and Merleau-Ponty's later work. As already mentioned, the completion of *Phenomenology of Percep-*

*tion* had left Merleau-Ponty with an acute awareness of its limitations: The identification of the body with sensory perception had resulted in a philosophy that was unable to account for ideational meaning beyond perceptual meaning—that is, for the full range of social, cultural, and intellectual life. His project after *Phenomenology of Perception* was to develop an account that would allow a passage to the conceptual world, without severing it from the world of perception. If his earlier approach emphasizes incarnated meaning as a positive layer to return to, his later work emphasizes instead the expansive and expressive dynamic of flesh, which does not stop at embodied perception, but extends into and comprises intellectual life (Saint Aubert, 2008, 10, 14). This expansive and dynamic notion of flesh presents a very different notion of mediation, not in its representational sense, where the medium is conceived as a vehicle for a pre-given meaning. Flesh cannot be parsed into a material and an ideational aspect (say, like signifier and signified in semiological structuralism), it is both at once. Drawing on some key quotations of Merleau-Ponty, we have coined the term the "measuring body" to emphasize the "in-each-otherness" (*Ineinander*) of the material and ideational aspects of mediation (Hoel and Carusi, forthcoming). To quote but one example of these:

> This means that instead of a science of the world by relations contemplated from the outside (relations of space, for example), the body is the measurement of the world. I am open to the world because I am *within* my body. But how do I have a sort of commonality with this mass of matter?—Precisely because it is *not* a mass of matter, it is rather a *standard of things*. (Merleau-Ponty, 2003, 217 emphasis in the text)

It is precisely by being a "standard of things" that flesh is neither a thing nor an idea (Merleau-Ponty, 1968, 152). As a "measuring" body, the body intervenes into the world; injecting a standard that serves as a point of orientation and institutes a *style* of seeing and accessing the world. The body actively measures; but the formative and constitutive activity is bidirectional: How it measures is inflected by what it measures; neither the measuring body nor the things it measures are mere things, but are immanently ideational as well. They bring to the measuring situation their own "logic," or, if one prefers, their own "agency." As such, the measuring body is emblematic of flesh. Through the notion of the measuring body, we bring together the scattered clues left by Merleau-Ponty into an account that accentuates the expansive dynamic of the flesh, and reconfigures the perceiving body into a symbolically and technologically distributed measuring body, bringing to the fore the ontological import of symbolisms and tools, which, each in their own way, serve as "measures of being" (Merleau-Ponty, 1973, 124).

There are different ways to trace the interconnectedness of the natural, symbolic, and technological in Merleau-Ponty's later work, but here we shall do so through noting some of his uses of the term "circuit."[7] We have already noted that flesh as formative medium indicates its in-betweenness (being in the middle; instituting a specific environment). The in-betweenness is also accentuated in Merleau-Ponty's use of the notion of the circuit. The idea of the circuit illuminates the "logos of perception" or "natural symbolism," as it does other symbolisms and tools or technologies. A pivotal section of his lecture notes on the concepts of nature deals with animality, and sees Merleau-Ponty engaging with the biological theories prominent at this time—particularly those that are attempting to give a non-causal, non-mechanistic account of animal development and behaviour. Prominent among these is Jakob von Uexküll, who stressed the dynamic interactions between the organism and its environment (*Umwelt*). The notion of the *Umwelt* brings out the reciprocal relation between nature that has created the organism, and the organism that creates nature (its *Umwelt*). The organism is in a circuit with the environment *with which* and *within which* it interacts. In their ongoing interaction[8] and intertwinement, both organism and milieu are mutually constituted.

In the last section of *Nature*, Merleau-Ponty works with these insights from biology, trying to see to what extent they can serve as ways of rethinking the ontology of perception. Merleau-Ponty, like Uexküll, stresses a continuity between the natural and the cultural or symbolic order. In Merleau-Ponty's later thinking life is conceived in terms of a productive negativity within being (the invisible scaffolding of the visible), whereby the living organism diverges from its environment in an ongoing process of differentiation and articulation. This role of differentiation and articulation through divergence is enacted by the measuring body, which distributes the environment according to its measure, while at the same time being shaped in response to that environment. In its measured relation to its environment, the living organism is not devoid of intelligibility: "animality is the logos of the sensible world" (2003, 166); it is "an apparatus of organizing perspectives" (2003, 166). Merleau-Ponty refers to this process of measured and mutual differentiation as a "natural symbolism" (2003, 212). He also speaks of perception as a "language before language" (2003, 219). This close interlinking of language and perception means that both have the capacities that Merleau-Ponty associates with symbolic systems. In virtue of being an "apparatus of organizing perspectives," we can go so far as to say that the perceiving body is a symbolism or language. Merleau-Ponty explains: "An organ of the moving senses (the eye, the hand) is already a language because it is an interrogation (movement) and a response [...], speaking and understanding" (2003, 211). Language proper, then, both continues and displaces the differentiating power of the body, due to the way that it "reproduces the perceptual structure at another level" (2003, 213).

Earlier we noted that there is, for Merleau-Ponty, a very close connection between language, technologies, and art, which are all seen as being embedded in the circuit between seeing and seen. In "Eye and Mind," for instance, he uses the example of the mirror: "Like all other technical objects, such as tools and signs, the mirror has sprung up along the open circuit between the seeing and the visible body" (1993,129). The circuit that interrelates and co-constitutes seeing and seen, knowing and known, is a *mediated* circuit—which is to say that it is not *one* unitary circuit of *one* modality; the measuring body is a shifting matrix that is at once perceptual, symbolic, and technological. The world opened and accessed in and through the measuring body is multimodal and multidimensional. Thus, in humans, the circuit is always already inflected by "artificial" symbolisms. Importantly these also include mathematics and algorithms, to which there are also several references throughout *Nature*. Some of these suggestive remarks point to a deployment of mathematics and statistics as a way of overcoming atomistic studies of isolated phenomena, and working instead towards "phenomena-envelopes" (2003, 247).

Merleau-Ponty also seems intent to reframe mathematics and algorithms and include them among the forms of expression. For example, he writes: "The modern evolution of mathematics which gets over the dilemma of quality or quantity. Theory of mathematics and of the algorithm to be made a variant of language" (2003, 313). The inclusion of mathematics and algorithms among symbolic systems, and therefore, we argue, in the measuring body, is one of the most significant moves that makes Merleau-Ponty's incipient philosophy of technology capable of doing justice to computational technologies. Elsewhere, we have attempted to draw out the implications of this approach for understanding the workings and functions of computational technologies in science, and their role in generating visualizations for computational biology and neuroscience, respectively (Carusi and Hoel, 2014a; 2014b).

Thus, by coining the term "measuring body," we further develop Merleau-Ponty's idea of the body as an "apparatus of organizing perspectives," by emphasizing even more strongly than does Merleau-Ponty the mediated nature of knowledge and being, and with that, the indispensable roles played by "artificial" symbolisms and technologies. The further emphasis on mediation also means that the interrogated apparatus is never given once and for all, but capable of being modified by the instruments, tools, and symbols at its disposal. Artificial symbolisms and technologies, in other words, are understood to have the power to transform the body's organizing logic and hence its relationship to its environment. By being injected into the circuit, instruments, tools, and symbols productively displace the inner firmament of our world, giving rise to new and *distributed* ways of interrogating the world; by being incrusted in the flesh, they allow new perceptions and actions and give rise to new kinds of entities—hence our contention about the ontological import of symbolisms and technologies.[9]

In this short overview we have highlighted a few lines of thought about technologies that point towards a Merleau-Pontian inspired form of postphenomenology, which, as we mentioned in the introduction, intersects with the ongoing development of postphenomenology on several points. The understanding of technologies that emerges from this reading of Merleau-Ponty accentuates the transformative and productive roles of technologies as they hook into the human perceptual system and form a distributed interrogating system—a measuring body.

We end with a discussion of a point on which there is an apparent divergence, that is, regarding Merleau-Ponty's criticism of science. This is remarked upon by Ihde and Selinger (2004), who see Merleau-Ponty as exemplifying the classical phenomenological critique of science and its technologies as alienating human experience. Our reading of Merleau-Ponty stresses instead that it is a certain *thinking* about science that is criticized by Merleau-Ponty and not its *practice*, which instead, has much in common with the forms of expressivity found in painting (see for example the authors' discussion of the use of visualizations in neuroscience, Carusi and Hoel, 2014b). The target of Merleau-Ponty's critique is often the rhetoric surrounding science rather than science itself; he is particularly dismissive of sciences as they become ideologies, and their terms of reference become hegemonic. This is the core of his criticism of cybernetics. For all that, at many points, it is clear that Merleau-Ponty is open to thinking in terms of cybernetics and computation (not least in his very use of the notion of the circuit).[10] For example, both in *Nature* (2003, 275) and in "Eye and Mind" (1993,127), he refers to the computational nature of vision and perception. However, he is critical of cybernetics (or at least its rhetoric) for having cast issues of perception and computation in machine terms, because he sees this as allied with mechanistic and causal thinking—that is, with thinking that is still caught in externalist relations between vision, technologies and things. Merleau-Ponty would thus hold cybernetics to have misunderstood itself precisely to the extent that it does not recognize itself as embedded in relations of "in-each-otherness" that characterize the circuit. Thus, far from criticizing technoscience from a distance, Merleau-Ponty exemplifies a mode of engagement with technoscience that bids us to attend to the criss-crossing mediations that together constitute scientific domains. By following through on this example, we can go some way to realizing the potential of Merleau-Ponty's thinking to reconfigure technology's role in knowledge and being. One route to doing so is in the form of the measuring body as a distributed perceptual, symbolic, and technological body that forms a particular form of apparatus for targeting, engaging with, interrogating, and forming the world in which that body operates.

## NOTES

1. Another major work by Merleau-Ponty on embodiment is *The Structure of Behavior* (1963).

2. The incomplete manuscript and working notes of the book that Merleau-Ponty was working on when he died.

3. *Nature* (2003) is a set of lecture notes by Merleau-Ponty and by his students between 1956 and 1960. The last part of the work ("The Concept of Nature 1959–1960: Nature and Logos: The Human Body") consists of Merleau-Ponty's own, often sketchy, notes.

4. For a critical discussion of the window metaphor and its implications for understanding the workings and functions of brain images, see Carusi and Hoel, 2014b.

5. See *The Visible and the Invisible* (1968, 176, 200) for Merleau-Ponty's own account of the shortcomings of *Phenomenology of Perception*.

6. For a more detailed discussion of the world-making capacities of flesh, see Hoel and Carusi, forthcoming.

7. See Hoel and Carusi, forthcoming, for a discussion of a wider range of terms associated with this.

8. With Barad (2007) we could even say "intra-action."

9. For a more detailed account of the measuring body and the positive account it provides of the transformative roles of symbolisms and technologies, see Hoel and Carusi, forthcoming.

10. Cybernetics was a prominent idea in intellectual circles in France during the 1950s (Mindell et al., 2003), and the idea of feedback loops is clearly evident in the circuit. Several of Merleau-Ponty's contemporaries also use the notion of circuit, such as Georges Canguilhem and Gilbert Simondon.

## REFERENCES

Bannon, B. E. (2011). "Flesh and Nature: Understanding Merleau-Ponty's Relational Ontology." *Research in Phenomenology* 41: 327–357.

Barad, K. (2007). *Meeting the Universe Halfway: Quantum Physics and the Entanglement of Matter and Meaning.* Durham, NC and London: Duke University Press.

Barbaras, R. (2004). *The Being of the Phenomenon: Merleau-Ponty's Ontology*. Bloomington: Indiana University Press.

Buchanan, B. (2008). *Onto-Ethologies: The Animal Environments of Uexküll, Heidegger, Merleau-Ponty, and Deleuze*. New York: SUNY Press.

Canguilhem, G. (2001). "The Living and its Milieu." Translated by J. Savage. *The Grey Room.* 3: 6–31.

Carbone, M. (2004). *The Thinking of the Sensible: Merleau-Ponty's A-Philosophy*. Evanston, IL: Northwestern University Press.

Carusi, A., and A. S. Hoel. (2014a). "Toward a New Ontology of Scientific Vision." In C. Coopmans, J. Vertesi, M. Lynch, and S. Woolgar (eds.), *Representation in Scientific Practice Revisited*. Cambridge, MA: MIT Press, 201–221.

Carusi, A., and A. S. Hoel. (2014b). "Brains, Windows and Coordinate Systems." In A. Carusi, A.S. Hoel, T. Webmoor, and S. Woolgar (eds.), *Visualisation in the Age of Computerisation*. London and New York: Routledge, 145–169.

Cataldi, S. L., and W. S. Hamrick. (eds.). (2007). *Merleau-Ponty and Environmental Philosophy: Dwelling on the Landscapes of Thought*. New York: SUNY Press.

Haraway, D. (1988). "Situated Knowledges: The Science Question in Feminism and the Privilege of Partial Perspective." *Feminist Studies* 14 (3): 575–599.

Hoel, A. S., and A. Carusi. (forthcoming). *The Measuring Body.*

Ihde, D. (1990). *Technology and the Lifeworld: From Garden to Earth*. Bloomington and Indianapolis: Indiana University Press.

Ihde, D. (1999). *Expanding Hermeneutics: Visualism in Science*. Evanston, IL: Northwestern University Press.

Ihde, D. (2002). *Bodies in Technology.* Minneapolis: University of Minnesota Press.
Ihde, D., and E. Selinger. (2004). "Merleau-Ponty and Epistemology Engines." *Human Studies*. 27: 361–376.
Latour, B. (1999). *Pandora's Hope: Essays on the Reality of Science Studies.* Cambridge, MA: Harvard University Press.
Latour, B. (2003). "Interview with Bruno Latour." In D. Ihde and E. Selinger (eds.), *Chasing Technoscience: Matrix for Materiality.* Bloomington and Indianapolis: Indiana University Press, 15–26.
Low, D. (2000). *Merleau-Ponty's Last Vision: A Proposal for the Completion of The Visible and The Invisible*. Evanston, IL: Northwestern University Press.
Merleau-Ponty, M. (1962)[1945] *Phenomenology of Perception*. Translated by C. Smith. London: Routledge & Kegan Paul.
Merleau-Ponty, M. (1963)[1942] *The Structure of Behavior*. Translated by Alden L. Fisher. Boston: Beacon Books.
Merleau-Ponty, M. (1968)[1964]. *The Visible and the Invisible*. Translated by A. Lingis. Evanston, IL: Northwestern University Press.
Merleau-Ponty, M. (1973)[1969]. *The Prose of the World*. Translated by J. O'Neill. Evanston, IL: Northwestern University Press.
Merleau-Ponty, M. (1993)[1961]. "Eye and Mind." In G. A. Johnson (ed.), *The Merleau-Ponty Aesthetics Reader: Philosophy and Painting*. Evanston, IL: Northwestern University Press, 121–149.
Merleau-Ponty, M. (2003)[1995]. *Nature: Course Notes from the Collège De France*. Compiled and with notes by D. Séglard; translated by R. Vallier. Evanston: Northwestern University Press.
Merleau-Ponty, M. (2008). "La Nature ou le monde du silence. Pages d'introduction." In Saint Aubert (ed.), *Maurice Merleau-Ponty*. Paris: Hermann, 41–53.
Merleau-Ponty, M. (2011). *Le Monde Sensible et le Monde de L'expression: Cours au Collège de France, Notes 1953.* E. de Saint Aubert (ed.). Genève: Metis Presses.
Mindell, D., J. Segal, and S. Gerovith. (2003) "Cybernetics and Information Theory in the United States, France and the Soviet Union." In M. Walker (ed.), *Science and Ideology: A Comparative History*. London: Routledge, 66–95.
De Saint Aubert, E. (ed.) (2008). *Maurice Merleau-Ponty*. Paris: Hermann.
Selinger, E. (ed.) (2006). *Postphenomenology: A Critical Companion to Ihde*. Albany: SUNY Press.
Sobchack, V. (2004). *Carnal Thoughts: Embodiment and Moving Image Culture*. Berkeley: University of California Press.
Toadvine, T. (2009). *Merleau-Ponty's Philosophy of Nature*. Evanston, IL: Northwestern University Press.
Verbeek, P-P. (2008). "Cyborg Intentionality: Rethinking the Phenomenology of Human-Technology Relations." *Phenomenology and Cognitive Science*. 7: 387–395.

## *Chapter Five*

# Movies and Bodies

## *Variations of the Embodied Self in Science-Fiction Techno-fantasies*

Marie-Christine Nizzi

The role of the body in defining the human self has been discussed since Ancient time, starting with Plato's claim that the body was a jail for the self (*Phaedo, 66b-e*), conceived as an immortal soul. Later, Descartes, although famously known for carving the mind-body dualism, tried his best to account for the role of this specific body that is mine in the identity of real men, understood as the deep union of a mind and a body (*Meditations*, 1641). The current state of science—including cloning, stem cells, and synthetic prosthetics—is a powerful source of revival for this issue in media. Is the body a mere container for the self or should we rather say that our thoughts, feelings and personality are the direct product of our physiological disposition? In analytic philosophy, science fiction or thought experiments are said to provide us with informative data about our beliefs on the true nature of identity (Parfit, 1984). Philosophers have thus imagined fictional cases like the body duplication or clone case (Nozick, 1981), the teletransportation machine or the divided-mind victim (Parfit, 1984), the exchange-of-bodies machine (Williams, 2003), or the brain-transplant victim (Shoemaker, 1963). I shall argue that science-fiction movies give an even more vivid and pre-theoretical experience of such concerns, and may thus be analyzed in terms of a mass experiments about folk-beliefs on body and the self.

As we consider multiple fictional scenarios for the future of human technology and its impact on our conception of our embodied self, we shall ask the following question: Is the self multistable? Let us distinguish immediately this question from another question: we are not asking here about multiple realizations. The multiple realization discussion assumes that there is such a thing as a specific mental state—say pain—and that this identical mental state can be produced by functionally different systems (Putnam, 1967). The whole point of the argument is that different systems realize the exact same construct (Fodor, 1974; Block, 1978; Endicott, 1993; Horgan, 1993). On the contrary, in the multistability discussion (Ihde, 1977), we are not assuming the identity of the construct. Rather, we focus on how changes in the techno-scientific context can modify the perception, interpretation, and definition of an initial object, in our case: the self. In other words, we are interested in how our concept of self might be altered as a function of a new technological framework.

Technology has massively transformed the human way of life over the past century. Not only has our everyday life become largely automatic (washing machine, automated vacuum, microwave, lawn-mower, snow-blower, remote controls devices, voice activated commands, etc.), the fundamental modalities of our interaction with others and the world have also been metamorphosed beyond recognition. Think about the social realm where internet has made collective memory virtually infinite and has demultiplied access to knowledge and information. Our relationship to time has been altered as news from across the world can be relayed in real time by radio, TV, cable, satellite, and internet. As things became more accessible, so did people, and the notion of privacy gained a new place in the spotlight as our electronic life made the line between private and public less intuitive (land-lines, then cell phones with GPS, real time social media like Twitter, international privacy scandals like the NSA in 2013). Similarly, our relation to space seems to have shrunken with the generalized use of cars, trains, airplanes, or video conference. Our relation to our environment has been drastically revised in some of its most basic features as we gained independence from natural light (electricity), temperature (heating, AC), and lack of water (irrigation).

More to our point, our relation to ourselves as embodied beings has been completely reshaped by technological progress. Our identity is now linked to finger prints, dental records, and DNA, yet it seems like the body has never been so technology dependent. What about the body's limits in strength, speed, or accuracy? Industrial robots now handle most of the heavy lifting, demolishing, assembling with precision and speed, sewing, planting. What about the body's fragility to pathological agents? Modern medicine has cured and sometime eradicated pandemics that used to cause millions of deaths every year. Babies who would not survive premature birth now receive intensive care. Unconscious patients receive automated life support. Surgeons can rely on medical robots to assist them in high precision acts and reduce the recovery time for invasive procedures. What about the irreplaceability of our body? The limits to organ transplants are pushed back every year. What about the biological nature of our body? We now have artificial hips and knees, pacemakers, synthetic prosthetics, even 3D-printed plastic skulls![1] Technology has even found a way to read others' minds, challenging the ultimate frontier of the private self by detecting consciousness in vegetative patients (Owen et al., 2006) and translating trained cerebral activity in paralyzed patients into spelling, voice synthesis, or motor actions via brain-computer interfaces (Oken et al., 2013). It is in such a context that we want to ask how technology impacts our concepts of self and embodiment.

Postphenomenology as defined by Ihde (2009) aims to integrate last century's techno-scientific progress by giving a major empirical twist to the traditional interest of phenomenology for the qualitative experience of being a subject. Subjectivity, in this view, must be studied in relation to its techno-scientific context. For instance, one can use technological mediation to gain access to the first-person feelings and decisions of non-communicative patients (Nizzi et al., 2012). One can also investigate the impact of mobile communication devices on our subjective experience of embodiment during driving (Rosenberger, 2014; Wellner, 2014). Those are only a few examples of investigations on how our concept of embodied self is transformed by the use of technology.

In this chapter, we adopt an epistemological strategy inspired by the philosophical tradition of thought experiments. Interestingly enough, this tradition is deeply rooted in the philosophy of mind and in the question of the role of the body in defining the human self. Locke introduced a famous thought experiment in his *Essay Concerning Human Understanding* when he suggested the reader to consider the case of a prince who would wake up in the body of a cobbler, to support his theory of psychological continuity. More recently, Parfit (1984; 2003) justifies the use of such fictions: "By considering these cases, we discover what we believe to be involved in our own continued existence, or what it is that makes us now and ourselves next year the same people. We discover our beliefs about the nature of personal identity over time. [. . .] Our beliefs are revealed most clearly when we consider imaginary cases." In the same perspective, we believe that science-fiction blockbusters seen by millions of spectators throughout the world gain their success by appealing to widely spread concerns in the general public. By studying how these movies question the role of the body in different hypothetical technological worlds, we are extending Parfit's strategy to identify beliefs held by a large audience of non-philosophers, thus extending the social bearing of the conclusions we can draw.

Let us consider the latest movie: *Captain America, the Winter Soldier* by Joseph and Anthony Russo (2014). Hopefully, by the time this gets published, it won't constitute any spoiler but in case it does, let us just say that one of the corrupted scientists who was believed to be long dead, turns out to exist now as a computer. Suffering from a fatal condition, he reveals that his body was lost but his "mind was worth saving" so it was transferred into a high memory set of connected hard drives. As we will see, this is merely one of the many ways in which humanity plays with the idea of surviving one's own biological body. In this case more so than in others, however, it was clearly spelled out that the body did not matter, only the mind did. Time and again, we find this dualist temptation to consider the body a mere recipient for the self rather than being its very mode of existence. As Ricoeur says, in such fictions, "the corporeal and terrestrial condition itself becomes a mere variable, a contingent variable."[2] Because we focus here on techno-scientific *fictions*, the question in our view cannot be "are these different scenarios preserving the self?" but necessarily becomes "what can we learn from our own fantasies about what makes us human?" and, more specifically, "how determinant do we believe our unique biological mortal body to be for the self in a world in constant techno-scientific progress?"

I want to claim that our body determines us absolutely *in the present*. That is, at any given time, my motor capacities, for instance, depend on my body and I am not at liberty to suddenly walk through solid walls by my sole desire to do so. It is not to say that human embodiment is defined by a static set of properties independent from its techno-scientific context. The human embodied self has already undergone considerable change throughout the history of evolution. With the help of technology, we can now fly when it was a radically inaccessible set of motions for humans before aviation. We can talk to each other while being separated by a great physical distance. Yet no one seems to think that we are less human for that. We can create lightning at will in controlled conditions and start avalanches in a way that would seem totally beyond the human range of possibilities to a society with less advanced technology. Certain eyesight impairments can be fixed by laser! Did that not sound like sci-fi just a few decades ago?

My point here is that we can only address the question of a multistable self by realizing first how our very notion of self is *dynamic*. I said that the body determines us absolutely: it does, in a given techno-scientific framework. The set of possibilities open to embodied humans is mediated by the technological context they live in. At present, on Earth, one could argue that there are several types of embodied selves. For some humans, it is possible to survive cardiac failure if a transplant is done in time, for others this very possibility does not even exist. The limits, the fragility, the uniqueness and the exclusive biological nature of the human body are already a fragmented reality depending on the technology one has access to. Yet science-fiction and thought experiments always seem to assume that there is something like a decisive step ahead that will break us apart from our human "original" nature. This only holds in a static ontology of the self and the body, one where being human refers to a fictive unchanging state outside of evolution and time. For those of us who live in a world where time passes, technology has broadened the spectrum of human possibilities from the very moment a stone got picked up and used as a tool.

The main lesson of adopting a dynamic framework to think about the self is that it allows us to look back in time to extract regularities, patterns, and ultimately principles that can shed light on the future we question in sci-fi movies. I want to point out two of these principles that describe how our experience and concept of the embodied self change following techno-scientific progress: (1) functional pragmatism, and (2) the consecutive evolution of the norm.

By *functional pragmatism*, I mean a fundamentally adaptive feature of humans by which we adopt new technologies and update the set of our possibilities in a continuous, forward-looking, smooth motion. The Google glasses are still a prototype. Within five years, they will most likely be a generalized accessory in industrialized countries. If it makes our lives easier, if we have pragmatic reasons to adopt it, then we do. Individual cars, cell phones, the internet, GPS, synchronized online calendars, credit cards, experimental treatments—did you really ask yourself if you were leaving an essential part of her human self behind when you gave your grandmother a cell phone so that she could call someone if she needed assistance? And now you are considering giving her that other little device to wear around her neck that will automatically call the emergency service in case she falls. Yet the human body is not by itself connected to a shared information network that makes such awareness possible: by adopting this lifeguard system, you change an elementary feature of her embodiment by allowing an online monitoring of her physical condition. Yet most people use this device as another tool at their disposal, just another technical means to an end. That is functional pragmatism. It predicts that we will integrate growing amounts of technology in our everyday life to serve various functions and it describes how we have done it over the past century more actively than ever before without feeling like we had become non human for doing so.

By the consecutive evolution of the norm, I refer to the update of social expectations following techno-scientific progress. Did you ever get frustrated because someone did not return your calls and emails over the weekend? Two generations ago, no one would have expected you to respond to a work-related question over the weekend. Did you have to set up an automatic message so that people who email you would understand your lack of response because you are "away from your desk until x date" and you will—of course—respond "ASAP"? What urgency! There was a time when getting an answer to a letter could take several months if it had to travel around the world. Yet people complain if they don't get it immediately, now that it is possible. We used to find the first modems fast, but now they would give us a heart attack because we got used to much faster connections. . . . And the list goes on. *Expectations conform to possibilities*.

And that is one aspect that we will find is lacking in many sci-fi movies: because they make sudden jumps into an imagined future, they skip the duration that enables adaptation. It is the elimination of this "time passing" variable that makes it novel and shocking and marginally scary to us, the spectators from present time. However, in the real world, things take time to diffuse throughout society. Cell phones were once restricted to an elite and GPS was first developed for military purposes. By the time they become available to the general public, the techno-scientific frontier has already moved forward so that the same technologies now seem only luxurious before they finally become just normal. But it is not the cell phone that has become normal. It is the massive social framework constituting "the norm" that has moved forward to integrate the cellular technology. Our expectations of immediate accessibility, of social connection despite physical distance, have grown enough to make the generalized use of cell phones seem natural. Groups who reject the use of technology on the other hand now contrast more and more with the constantly updating norm and although we might conceive of their life style as more natural, the overwhelming choice of societies has been to go with technology. I voluntarily adopt a descriptive perspective here and not a prescriptive one. My aim is not to say where the true human embodied self is, if such a unified static notion even makes sense, but rather to notice that the exception has changed side. And that is, to me, the second factor that we will need to consider in our movie examples. The evolution of the norm determines what the exception is.

With that in mind, let us now explore several science-fiction block busters that target three core issues related to the embodied self in various futuristic techno-scientific worlds: the desire to expand the range of our capacities, the fear to lose ourselves in the process, and the question of what a disembodied human self would be like. We shall apply the happy ending criterion to evaluate whether each issue constitutes an utopist or dystopist fiction. According to this criterion, fictions that tend to be received by the general public as utopist will triumph in the movie while fictions feared as dystopist shall suffer the fate of the bad guy.

## EMBODIMENT AS A FRONTIER

In today's real world, humans are born as a unique, biological, mortal body. We don't choose our body nor can we exchange it at will. It has a determined range of possibilities, both in the sensorial dimension (we do not perceive ultra-violets) and in the motor dimension (we cannot fly on our own). Our body also determines our temporal extension (roughly one hundred years) and our relation to matter (we cannot pass through solid matter). In other words, the characteristics of the human body determine the modalities of existence of the human self. For all science can tell today, there is no human self without embodiment. Now, for science-fiction, that is an open play ground. By modulating each of the parameters we mentioned above, sci-fi movies explore how various technologies would transform the definition of our embodied self.

## *Avatar*: A New Set of Possibilities

The 2009 block buster movie by James Cameron, *Avatar*, presents us with a very interesting exploration of the possibilities open to human embodiment through technology. The movie takes place in 2154, at a time when humans have depleted Earth's natural resources and explore the galaxy to find more resources. The planet Pandora is inhabited by the Na'vi, ten-foot-tall, sapient humanoids who can breathe the local atmosphere toxic to humans. To explore the planet, scientists use Na'vi-human hybrids called "avatars." Jake Sully, a paraplegic former marine, becomes one of the explorers. Through a machine that synchronizes his brain activity with the avatar, Jake gains a full sensory-motor experience as from within the avatar itself.

Quite plausibly although very rapidly, the first time Jake tries out his avatar, the scene shows him learning by trial and error—meaning by breaking a great deal of equipment around him in the lab—how to adjust his body schema to his new body, control his motor actions, and make sense of the radically new proprioceptive features of having a tail. When patients who were born without an upper limb are equipped with a prosthetic arm for the first time, the motor and premotor cortices follow a similar learning curve to adjust the reaching and grabbing motions. However, as our technological level is still limited, the prosthetics don't yet provide the patients with as rich a sensorial experience as Jake seems to experience on screen through his feet pounding the dirt and his other senses receiving extraterrestrial odors, sounds, and sights for the first time. The possibilities opened by this projected body are even more salient that the human embodied Jake is paraplegic. The use of the avatar fixes his impairment and gives him motor possibilities back.

Na'vi do belong to science fiction. Yet the basic premises of allowing a paralyzed individual to control a second body by projecting their brain activity to the nervous system of that body have opened a new lead for research. Five years after the movie, scientists have managed to enable a monkey to control the movements of another anesthetized monkey through a "cortical–spinal neural prosthesis that employs neural activity recorded from premotor neurons to control limb movements in functionally paralyzed primate avatars" (Shanechi, Hu, and Williams, 2014). Such research could eventually lead to reconstituting targeting movement in paralyzed patients.

How would this technology impact our concept of embodied self? It suggests that there could be two bodies for a single individual. Granted, the motor control cannot be maintained in both bodies at the same time, so we are considering a case of fractioned embodiment rather than duplication. But the impact of such a technology could go beyond acting through another human body. The movie explores what it could feel like to project oneself in a different body, one that—as it turns out in this specific fiction—has additional capacities compared to the human body. Indeed, the scenario goes beyond giving Jake his legs back: the Na'vi have a way to connect to each other as well as to other creatures that creates a special bond. This experience is, by definition, not human. It is determined by the Na'vi body's own range of possibilities. Yet what would happen if one was to experience a full sensory-motor transfer with a body that had non-human possibilities? The movie at least disagrees with Nagel's famous article "What is it like to be a bat" (1974) and concludes that Jake can experience the Na'vi's connection, first with Toruk, a flying creature and then with Neytiri, his Na'vi mate. If that was in fact the case, based on our own experience or the proven possibility to have such experiences, our concept of embodied self would in all likelihood update to integrate new sensory-motor possibilities within its definition.

In the movie, Jake faces an ideological conflict that forces him to choose between his human embodiment and his Na'vi embodiment and he decides in favor of the Na'vi's existence. However, prior to choosing, he indeed describes his experience as an enlarged one rather than a completely separate set of experiences, thus maintaining the idea of a continuous self. The modalities of his embodiment still determine what kind of sensory-motor interactions he can have with his environment and only through his Na'vi body can he have non-human experiences. Yet it is not until the ideological conflict between us and them starts that those two types of embodiment become mutually exclusive. We see that, in contrast to the multiple realization discussion where the question would be "can the Na'vi's physiology produce the same mental states as human experience?" our question here is "can our concept of human embodied self extend so as to integrate non-human experiences?" But this might sound far-reaching. Let us talk about a fiction that is much closer to the current state of technology: the possibility to selectively enhance certain physical and cognitive functions through the use of drugs.

## *Limitless*: Human Enhancement on Prescription

In 2011 Neil Burger's movie, *Limitless*, the hero Eddie Morra takes an experimental drug, NZT-48, that immediately enhances his perceptive, cognitive and reasoning abilities. With enhanced memory, intelligence, and perception, he manages to outsmart the corporation trying to use him for their own purposes.

*Limitless* suggests less extreme a change than *Avatar*. The fiction is set in a world roughly identical to ours. The hero retains his human body and there is no off-world colonies populated with aliens. This time, the only change is one of degree in human capacities rather than one of nature. By all definitions, Eddie is still himself and he is still human. However, he is a better version of both, extending normal capacities beyond the average performance. This would be totally unremarkable if it was not drug induced, as intelligence and perception naturally come in a wide range across individuals.

The main concern that the movie touches upon is about the possible long term side effects of such an artificially induced state of hyper performance. The hero suffers from black-outs that leave him amnesic of his immediate past. Additionally, the drug poses the question of addiction. When the hero of *Avatar* chooses the Na'vi mode of embodiment, he knows it is a stable, viable option as Na'vi have lived for generations. On the contrary in *Limitless*, the drug is experimental. It is an artificial, manmade composite, the long-term side effects of which are unknown. The movie resolves both problems by suggesting that enhanced capacities enable the hero to find a solution to the pervasive effects of the drug. However, this approach points at a distinction regarding our concept of embodied self. On the one hand, it seems like developing normal human capacities beyond their natural extension does not affect our definition or our interpretation of what is a human embodied self. A change in degree of capacity is a much smaller gap than the change in nature required to integrate non-human experiences and it is very likely that we would readily expend our definition to integrate enhanced capacities. Yet on the other hand, there is a pragmatic concern limiting the enthusiasm for the fictional drug: our safety could be at stake. Here it is about another characteristic of our embodiment that we might be more reluctant to let go of: our body has evolved across millennia and most of the time without such an active part on the humans themselves. It is probable that there would be an initial resistance to the idea that humans can play with the brain's chemistry and get it all right. However, eventually, the functional pragmatism principle predicts that such enhancement technologies would be accepted and the movie accordingly has a happy ending.

Notice that the plot of *Limitless* focuses on a single individual. At most, a handful of people have access to the drug. Eddie's success as a writer, a business man, and later as a politician is at least in part due to the limited distribution of the drug, which gives him an advantage compared to non enhanced humans. What happens to our concept of human embodiment when such enhancements become available to more?

## *Elysium* and the New Margin

*Elysium* opens a very interesting reflection on the social impact of costly human enhancement technologies. In his 2013 movie, Neill Blomkamp depicts a divided world of the year 2154 on Earth. While the elite enjoy a protected life with perfect medicine on a technology-assisted habitat in orbit, Earth population lives in misery, administered by robots. After receiving a lethal dose of radiation during an accident, the hero Max Da Costa is willing to undergo an extensive and risky surgery adding to his human body a powered exoskeleton complete with neural implant. With his strength now matched with robots, he manages to invade Elysium and uses the program downloaded in his neural implant to override the main system, making all Earth inhabitants citizens of Elysium. As a consequence, a battery of medical ships departs to cure Earth inhabitants, thus giving an equal access to the perfect medical technology so far restricted to Elysium's citizens.

The issue of *who* has access to any considered technology is central to the question of a change in our collective concept of embodied self. A single individual with super-human capacities is a hero, a god, or a monster. A small group still only is an exception. But as soon as large enough a group has access to a given technology, there seems to be a switch that triggers the updating process of the norm. The movie emphasizes that the subsequent evolution of the norm happens even when the technology is not readily accessible to everyone. In this sense, the possibility prevails over the actual accessibility of the technology: as long as it is possible, it creates a new standard for what can be expected. No need for fiction here though: we experience this situation in the real world about a variety of technologies including medical technologies like transplants and joint replacements. Only a minority of Earth inhabitants has access to these technologies yet this simple fact makes it abnormal that the rest of the population doesn't.

A related phenomenon that the movie illustrates is the creation of a new margin, simultaneous to that of a new norm. When the citizens of Elysium can readily cure cancer and benefit from certain enhancements, all of a sudden, it creates a new category: even healthy humans appear to be somewhat less because of the possibility of enhancement. In other words, there are now three categories where there were only two: there are sick humans, healthy humans, and enhanced humans. The question is whether our concept of human embodiment would evolve to integrate this new possibility as its standard. Both the movie and our principle of norm adjustment suggest that it would. Let us take a real world example. Vaccines are a technology that prevents an individual from getting sick in the future or diminishes the symptoms so as to avoid death by the pathological agent involved. Suppose you lived in an area infested with a very aggressive disease for which there is no vaccine. In this situation, being healthy is the normal situation compared to which being sick is a deterioration. Now imagine we do have a vaccine. By its sheer existence, the vaccine transforms the healthy but not immune situation into a less than ideal situation. The expectation integrates the technological progress into a new norm. Now consider the cost of the vaccine and how only a fraction of the population can get access to it. Because of the norm updating principle, people who cannot afford the preventive medication now find themselves less fortunate by the simple fact that they cannot benefit from an existing enhancing possibility and the people who are sick when the vaccine is available are doubly unfortunate. This new margin phenomenon would happen regardless of the nature of the technology: it happens whenever enough individuals of a population have an exclusive access to a superior technology.

The fact that the enhanced human becomes the new norm rather than becoming an upper possibility while the healthy not enhanced human would remain the norm illustrates how the concept of embodied self would be modified by techno-scientific progress. Just like it is not the cell phone that has become normal but the set of expectations that has come to integrate it, it is not the enhancement technologies that have become normal, it is our concept of embodied self that has grown to include them in its definition.

## THE FEAR OF THE OTHER: HUMAN VERSUS MACHINE

### *Blade Runner* or the Blurry Border

After this series of utopias, let us consider another category of science-fiction movies that often do not have the happy ending signaling hope for technological progress but rather an ambivalent ending that resonates like a warning, echo of a fear: that of loosing what human embodiment specifically means. As computational theorists of mind used the analogy of a computer to understand and model the human brain, film makers have jumped at the possibilities for fiction, be it along the line of creating humanoid robots that would one day rebel in the name of their superior cognitive functions (cf. *I, robot* by Alex Proyas, 2004) or along the line of humans transferring their consciousness into computers for various purposes (survival in the face of a lethal condition, preservation across intergalactic travels, or security of the human body). These movies directly question what it means for humans to be embodied.

Here we start by calling upon the 1984 Riddley Scott's movie *Blade Runner*. The movie takes place in 2019. Corporations produce genetically engineered organic robots called replicants to perform dangerous tasks on off-world colonies. The replicants are visually indistinguishable from adult humans. Their presence is banned on Earth and the hero, Deckard, hunts them. A scene of particular interest for our topic shows Deckard administering a "Voight-Kampff" test designed to distinguish replicants from humans based on their physiological response to questions with high emotional valence. Deckard is questioning the assistant of the Tyrell Corporation's president, Rachel, an experimental replicant who believes herself human on the basis of implanted memories. As a consequence of her unique design, the test requires considerably more questions than usual to identify her as a robot. The hypothesis is that the fake memories (standard across several robots) provide Rachel with a more individualized sense of self and with an emotional reference. So which parameters of human embodiment does this fiction manipulate?

In the movie, not only do the replicants have biological human shaped bodies, they are also becoming emotionally and morally closer to humans. The final fight between the replicant Roy and the hero in which Roy saves Deckard rather than killing him after reflecting upon the morality of his past actions suggests as much. So do the emotional capacities of Rachel. And, at the limit, that seems to be the concern: if enhancements can be integrated easily into a new norm for our concept of embodied self and if these robots are by every characteristic but their generation biological replicas of self-conscious humans with enhanced capacities, then what makes us an embodied self and them not? More importantly, from a pragmatic and ethical point of view, can we still use them as we do with tools? The movie does not answer these questions. It only raises them as problematic. With the exception of Rachel, whose future is unknown, the replicants fail and die. They do not become the new norm and the concept of embodied self does not extend to them collectively. In this case, it seems that the decisive factor is that replicants are artificially engineered. Despite their body being biological, subject to sensations and emotions, and mortal like human bodies, their artificial generation seems to be what ultimately distinguishes them from human embodiment. That is not the only scenario fictions have envisioned to illustrate how technology could drive humans to lose the meaning of their embodied self.

## *Surrogates* and the Loss of Self

In his 2009 movie entitled *Surrogates*, Jonathan Mostow depicts a future in which people use remote-controlled androids called surrogates to live their lives from the safety of their homes. Surrogates can be damaged or destroyed without the user being hurt so that humans using them survive accidents and crime committed in the streets towards their surrogates. A neural connection gives the user control over their surrogate, whose appearance can be completely different than that of the real user. The hero, Tom Greer, is one of the users. After the death of their son, his wife refuses to interact with him as her human self and restricts their relationship to that of their surrogates.

This fiction raises several interesting points about the human embodied self. Are we willing to use surrogates to gain physical safety? Is our physical appearance determinant in defining our experienced self and our social identity? Would we lose contact with our real self after repeated use of the surrogate? The movie explores these questions and provides us with answers relevant to our interest for the potential impact of fictional technologies on our concept of embodied self.

The first issue concerns our readiness to adopt the surrogate technology. The movie presents it as a technological progress granting the users a safer life. In a way, the impact on our embodied self seems minor. Everybody remains their usual embodied self. The surrogate is merely a puppet used outside of the house in addition to our embodied self. Come the evening, the surrogates go to their charging stations and the human family meets up for dinner. Or not… Although the movie does not play with the idea of transferring the consciousness of the user into the surrogate to make the user immortal for instance, it emphasizes the possible impact of having even an accessory-robot self on top of one's own embodied self. First, individually, there might be reasons to favor a robot whose appearance we can change at will. But also, collectively, the surrogates follow the updating norm principle. When the hero happens to walk the streets in the flesh, he experiences fear, suddenly fully aware of his physiological vulnerability. As soon as a small number of people can afford surrogates, the police will need to match the capacities of these individuals and citizens might begin to feel even more unsafe as much stronger robots now walk the streets. Not surprisingly, the advertisement messages from the company manufacturing the surrogates target the need for safety of their consumers. Following what the functional pragmatism principle would predict, the surrogates are quickly adopted in the movie and using them becomes the new norm. Now let us see how this technology would impact our concept of embodied self.

The second issue addressed by the movie concerns the role of our physical appearance in our experienced and perceived self. An amusing scene shows Tom interrogating one of the company's officials, flanked by two good-looking assistants. He points out in a very eloquent way how different the actual person could be from its surrogate. This raises two points: one is a matter of identification of individuals who can afford to own several different-looking surrogates, the second is a matter of social impact of physical appearance. The first one displaces the identification factors away from the body—at least away from the surrogate's body. No DNA or fingerprint to find on a puppet. At best, there could be a serial number with ownership records. That would constitute a major change for our concept of embodied self. As of now, it is extremely difficult to escape one's identity because of our body. We cannot change body as we change shirts. However, if the identification demand shifted towards remote-controlled robots, then the social self, our personal identity, must be redefined through non-bodily criteria.

The second point is that manipulating one's appearance can serve social goals. Increasing desirability or hiding signs of aging are only two of the possible motivations. Seemingly superficial, these motives in fact get at the root of human embodiment. We don't choose our body, we cannot change it all for another one, and we can't stop aging yet. If technology were to enable us to change these three facts, then it would also reshape human embodiment at its core. Would the self still be defined as necessarily embodied? It depends on what we mean by embodied self. Is it merely having a body (in which case it could potentially be extended to having several bodies) or does it mean being one's own and unique body? But then again, already in the real world, our body naturally changes a lot over a lifetime and no one is thought to lose their self when they undergo plastic surgery. Heavy physical transformations are even showed on TV as beauty contests. In a different domain, sex change operations are possible despite their direct impact on the person's identity. Maybe then should we consider the possibility that there is just a difference of degree between the current real-world situation and a fiction in which we could change our physical appearance at will through the use of surrogates? Perhaps our concept of embodied self would then just evolve to include the possibility of several physical appearances, and that would become the new norm. Consequently, choosing to live without a surrogate would result in marginalization, as it is imagined in the movie.

The last issue is tackled mostly in two scenes: first when Tom confronts his wife's surrogates demanding to see "his wife" and insisting that the surrogate is not his wife but just a robot, and then during the final scene, when all surrogates have been destroyed and people start getting in the streets, looking confused and afraid. Would the adoption of the surrogates technology lead humans to lose themselves after a fashion? The movie suggests that it would: there is no happy ending for the surrogates. Humans wake up, as from a long dream, and look at each other for the first time in years. Social connections require some face-to-face contact, suggests the movie. When we place distance between our embodied self and multiple social covers only loosely related to our real self, we not only lose contact with others, we also lose contact with ourselves. No one on one relationship can be trusted anymore as the person behind the surrogate could just be anyone. Trust is a surprising by-product of the quasi impossibility for humans to completely escape or change their physical self. When this unique relationship between body and self is lost, it seems that it is a fundamental component of the social fabric that vanishes. This is another way of illustrating the fear of blending into something that would not be us anymore. But this time not because we created robots undistinguishable from ourselves; just because changing our relationship to our body in too fundamental ways would also unravel the human social reality. Then should we conclude that no fiction offers an utopist thought about what a disembodied human self could be like?

## TOWARD A DISEMBODIED HUMAN?

### *Transcendence*

Throughout this chapter, we have investigated how techno-scientific fictions could shed light on our concept of embodied self. We have considered the fictive possibility of expending human capacities, of connecting ourselves to the body of a different species as well as to a robot. Now let us consider three types of fictions that directly address embodiment as a contingent variable in our experience of the self: first, by unleashing a mind online after its physical death; second by de-materializing the human body; and third by progressively, one step at a time, abolishing the time and space constraints that determine human existence in today's real world.

This movie hasn't been released, yet its trailer hits every point we have made so far: *Transcendence*, by Wally Pfister, follows the fate of Dr. Will Caster, a scientist working on creating a machine with both emotional and cognitive power. The motivation is again one of functional pragmatism: "early detection of cancer, saving lives." Yet, after activists opposing technology inflict him a fatal wound, Will's wife attempts to "save him by uploading his consciousness" into a computer. When Will reaches physical death, a message suddenly appears on the screen: "Is anyone there?" Will's wife believes this is proof that Will's consciousness is now existing in digital format and expressing itself on the screen. The entity says that it feels like its "mind has been set free" but soon it needs more power, the computer being too limited for a human mind. The entity asks her to put it online: on the distributed network supported by internet, it could get additional space and power. Soon, even Will's wife questions the identity of the entity, whom she doesn't recognize any more and it seems that the existence of the entity puts humanity as we know it at risk.

Why do I think this fiction will be among the dystopias? Because there is no functional pragmatism working for mankind here but on the contrary a risk of losing embodiment as a defining feature of our humanity with no turning back; because the movie is about a single mind striving for its own expansion at the expense of another form of life, thus making it the anomaly rather than the new norm; and because it is a one step process leading from a currently accepted body into a fully virtual existence rather than a progressive evolution. Although it is hard to be certain from the trailer only,[3] it seems like the movie doesn't suggest for all mankind to jump into a virtual existence. What is suggested is that a form of intelligence embodied in a distributed electronic network will fight for survival by expending onto the entire system. This cannot constitute a viable collective future. However, it points at another fictional consequence: when accessing the internet, when expending into a virtual distributed intelligence, the once human Will loses who he was as if his self was doomed to be altered beyond recognition in this radical change of embodiment. Not tied to one spatial location; not restricted to the human brain's limited computational power; not dependent on any human needs: by exchanging his body for the internet, Will loses himself because he now transcends the human condition. It is now a case of us versus the Other, the machine. In such a configuration, there can be no update of the norm. Instead, the entity is pushed outside rather than being integrated in the concept of embodied self and as the trailer suggests "whatever this is, it is not Will anymore." Here we conclude to the loss of self because the conditions of embodiment have nothing to do with the human condition anymore yet this distributed digital intelligence presumably still relies on the physical existence of the computers composing the network that supports its existence just as humans need their body. What happens in a fiction that renders any sort of material support contingent to the existence of a former embodied human self?

## *Watchmen:* Human Concerns?

In his 2009 *Watchmen* movie, Zack Snyder features a distant Dr. Manhattan. After his accident, the former scientist learns to control the molecular structure of his body thus becoming able to take a material embodied form at will, to manipulate the size of his body, and to recompose his organism from a fragmented molecular level suggesting he cannot die anymore. During the movie, he also demonstrates several non-human capacities including demultiplying his body into several autonomous yet connected agents, traveling through space without need for breathing, controlling matter in a telekinetic manner, and seemingly knowing the future.

One scene in particular is interesting in tying up his radically changed embodied existence with the question of his human self. While half of mankind faces annihilation, Dr. Manhattan's former girlfriend tries to convince him to intervene during a conversation on Mars, where he teleported her with him. During their conversation, his girlfriend points out that his knowing the future is "inhuman," which makes her uncomfortable. In response, he shows her how his interest has progressively shifted from Earth to Mars. This new world "means more to him" than Earth. She is "his only remaining link to humanity" and this status makes him wonder "why to save a world he no longer has any stake in." One very pragmatic way to look at humans' interest for planet Earth is that Earth is so far our one and only habitat. In one way, we collectively share with Earth an up-scaled but similar link as we share individually with our body in that there is no alternative for us yet. However, the movie suggests that this interest is ultimately tightly linked to the modalities of our embodied existence. Could we exist with no need for air or water, then maybe our bond to our environment would be shattered simultaneously to our bond to unique embodied conditions of existence. Indeed the human body has evolved to fit into a specific environment. What this movie points at is the potential large-scale change that making the body contingent to humans would produce in their relation to their entire environment.

We have considered with *Surrogates* the possibility that fundamental changes into our mode of embodiment would unravel human social reality but in that fiction, humans still were embodied selves. If one individual gained so much control over their body so as to manipulate their molecular composition at will, so as to abolish the uniqueness of spatial location that still constricts human embodiment in the real world, would we still believe them human? Because he is alone, Dr. Manhattan leaves Earth at the end of the movie, a single entity of its kind with no specific interest for humanity or Earth, to which he doesn't belong anymore. Yet if all of Earth inhabitants were to undergo the same transformation, would we not then revise our concept of embodied self? Instead of being tied to a single environment with its defined life conditions, would we free the self from a unique mode of embodiment to integrate the possibility to live in multiple environments offering multiple conditions of existence?

## *Gentle Seduction:* Human, Time, and Matter

My last stop on this itinerary across fictions is a short story, written by Marc Stiegler in 1989: "The Gentle Seduction."[4] This fantastic story follows a woman with no special drive for technology. She lives a simple life in the mountains, doing most things by herself. One day, as she turns eighty-two, she finds herself unable to shovel her driveway. In her hand, she holds a capsule that the pharmaceutical and nanotechnology industries have developed in the past few years:

> She opened her hand and looked at the capsule. It was not a pill to make her younger; that much her children had promised her. They knew she would reject such a thing out of hand. But the millions of tiny machines tucked inside the capsule would disperse throughout her body and repair every trace of damage to her bones. They would also rebuild her sagging muscle tissue. In short, the pill would cure her back and make the pain go away. (Stiegler, 1989)

She takes the capsule, then another one to fix her heart, and she starts hiking again. One day, another hiker saves her from an ice fall thanks to a technological headband connecting his mind to a network of computers to analyze information about the environment too subtle for human sensorial capacities.

She realizes how many deaths could be prevented by using this technology . . . and adopts it.

This is no place to rewrite the story, the merit of which is all due to its author. Instead, I will just suggest the road ahead. This story has a happy ending, yet a disembodied one. It is not the story of single individuals extracting themselves from humanity through technology, though. Nor is it the story of multiple cloning or of transferring one's consciousness into a computer. Instead, it is a very long journey through which humanity as a whole grows and gradually updates its norm of what being an embodied self means. It is the fiction that illustrates how small steps, smooth progress over an extended period of time spreading progressively throughout mankind, can take the meaning most of us have today for embodied self and makes it vary as a function of successive small technological changes. Our relation to time is the first one to be redefined as technology eliminates the effects of aging on the body. Then our relation to space expends beyond Earth, which eventually requires some alterations in our embodiment, to experience different possibilities opened by different environments. Eventually, even our relation to matter, to that very specific piece of matter that individuates us, becomes just one of several ways to embody one's self. Her self remains, her human body as well. Their relationship has just extended to simultaneously integrate other modalities of embodiment. And this fiction, by all we have seen so far (the functional pragmatism, the updating norm, the need for a collective evolution, the duration necessary for the integration of technological progress) is a story on the edge of tomorrow. I take this story to express what we hold to be an embodied self today and to show just how we would adjust its perception, its interpretation and its definition as the techno-scientific world we live in evolves.

## CONCLUSION

We started by asking whether the self is a multistable concept. Although we often forget it, our past can already answer this question. Yet science-fiction movies predict a decisive change in the future, this one technological move that will rip the evolution of mankind apart rather than develop it continuously. I have tried to argue that this fear only holds if one forgets that being human is a moving target: the very conditions of our existence, its meaning, change dialectically with our techno-scientific context. From analyzing some recent science-fiction work, we can identify broadly shared beliefs about what makes us human and what impact we think different types of technological changes to our current embodied condition would have on our concept of self. Several scenarios seem to lead to a loss of self by undermining core features of our embodiment, such as downloading one's consciousness into a computer, living through multiple artificial surrogates, or undergoing an exceptional event that would make only an extreme minority leap out of the mode of embodiment that would still govern the rest of humanity.

In the happy-ending, hopeful scenarios, on the contrary, certain conditions are met: the change is progressive, thus allowing successive updates of the norm; it serves functional pragmatic purposes, like saving lives, and it is accessible in principle to mankind as a whole, thus supporting the updating of the norm. In those scenarios, the perception, interpretation, and definition of an embodied self can vary to a great extent without losing its referential value. At any given time, it is still true that our body defines absolutely our possibilities, yet these possibilities vary directly as a function of technological progress. As a result, we can expect our concept of self to behave as a dynamic, dialectic, and, in a word, *multistable* reality of human experience.

## NOTES

1. http://www.nbcnews.com/science/science-news/medical-first-3-d-printed-skull-successfully-implanted-woman-n65576
2. See (Ricoeur, 1992, 151), as cited in (Ihde, 1998, 104).
3. I have had the chance to add this note right after the movie was released. Two points need to be updated in the light of the full movie (spoilers ahead): on the one hand, there seems to be at least a possibility for re-embodiment after taking on an electronic mode of existence and on the other hand, there seems to be a possibility for several persons to co-exist in this modality, thus extending it beyond a single individual survival's strategy. However, even then, the theory of mind question remains: how can we be sure that an entity is who it claims to be once it is possible for it to render its embodiment contingent? On that question, I refer the reader to the movie!
4. See http://www.skyhunter.com/marcs/GentleSeduction.html .

## REFERENCES

Block, N. (1978). "Troubles With Functionalism." In C. W. Savage (ed.), *Perception and Cognition: Issues in the Foundations of Psychology*. Minnesota Studies in the Philosophy of Science, vol. 9. Minneapolis: University of Minnesota Press, 261–325.

Descartes, R. (1996). *Meditations on First Philosophy*. Cambridge University Press.

Endicott, R. (1993). "Species-Specific Properties and More Narrow Reductive Strategies." *Erkenntnis* 38: 303–321.

Fodor, J. (1974). "Special Sciences: Or the Disunity of Science as a Working Hypothesis." *Synthese* 28: 97–115.

Horgan, T. (1993). "Nonreductive Materialism and the Explanatory Autonomy of Psychology." In S. Wagner and R. Warner (eds.), *Naturalism: A Critical Appraisal*. Notre Dame, IN: University of Notre Dame Press, 295–320.

Ihde, D. (1977). *Experimental Phenomenology*. Albany, NY: SUNY Press.

Ihde, D. (1998). *Expanding Hermeneutics: Visualism in Science*. Evanston, IL: Northwestern University Press.

Ihde, D. (2009). *Postphenomenology and Technoscience: The Peking University Lectures*. Albany, NY: SUNY Press.

Locke, J. (1689). *An Essay Concerning Human Understanding,* Chapter XXVII on "Identity and Diversity." Edited by R. Woolhouse, 1997. London: Penguin Books.

Nagel, T. (1974). "What Is It Like To Be a Bat?" *Philosophical Review* LXXXIII. 4: 435–50.
Nizzi, M.-C., A. Demertzi, O. Gosseries, M.-A. Bruno, F. Jouen, and S. Laureys. (2012). "From Armchair to Wheelchair: How Patients with a Locked-in Syndrome Integrate Bodily Changes in Experienced Identity." *Consciousness and Cognition* 21 (1): 431–437.
Nozick, R. (1981). *Philosophical Explanations*. Cambridge: Harvard University Press.
Oken, B., U. Orhan, B. Roark, D. Erdogmus, A. Fowler, A. Mooney, B. Peters, M. Miller, and M. Fried-Oken. (in press). "Brain–Computer Interface With Language Model–Electroencephalography Fusion for Locked-In Syndrome." *Neurorehabilitation and Neural Repair*.
Owen, A., M. Coleman, M. Boly, M. Davis, S. Laureys, and J. Pickard. (2006). "Detecting Awareness in the Vegetative State." *Science* 313 (5792): 1402.
Parfit, D. (1984). *Reasons and Persons*. New York: Oxford University Press.
Parfit, D. (2003). "Why Our Identity is Not What Matters." In R. Martin and J. Barresi (eds.), *Personal Identity*. Malden: Blackwell Publishing, 115–143.
Plato. (1966). *Plato in Twelve Volumes*, vol. 1. Translated by Harold North Fowler. Cambridge: Harvard University Press.
Putnam, H. (1967). "Psychological Predicates." In W. H. Capitan and D. D. Merrill (eds.), *Art, Mind, and Religion*. Pittsburgh: University of Pittsburgh Press, 37–48.
Ricoeur, P. (1992). *Oneself as Another*. Chicago: University of Chicago Press.
Rosenberger, R. (2014b). "The Phenomenological Case for Stricter Regulation of Cell Phones and Driving." *Techné: Research in Philosophy & Technology* 18 (1/2): 20–47.
Shanechi, M., R. Hu, and Z. M. Williams. (2014). "A Cortical-spinal Prosthesis for Targeted Limb Movement in Paralyzed Primate Avatars." *Nature Communications* 5: 3237.
Shoemaker, S. (1963). *Self-knowledge and Self-Identity*. Ithaca, NY: Cornell University Press.
Stiegler, M. (1989). "The Gentle Seduction." *Analog Magazine*.
Wellner, G. (2014). "Multi-Attention and the Horcrux Logic: Justifications for Talking on the Phone While Driving." *Techné: Research in Philosophy & Technology*. 18 (1/2): 48–73.
Williams, B. (2003). "The Self and the Future." In R. Martin and J. Barresi (eds.), *Personal Identity*. Oxford: Blackwell.

## *Chapter Six*

# Bodies as Technology

## *How Can Postphenomenologists Deal with the Matter of Human Technique?*

Fernando Secomandi

The alarm rings. As I turn to the side table to switch it off, the bed feels strangely not mine. Nor is this room my own! Oh yes, I am lodged in a hotel, for the conference. While still lying in bed, I check for incoming messages on the iPhone. There is no news from work. Then, I get up and into the shower, dress myself, and head outside the room towards the elevator hallway where other people wait. Most of them wear nametags, like I do. Riding the elevator down, I glance at one of the nametags. This is not a name that I recognize. I look at the person's face. We exchange smiles of complicity. At the entrance to the room where breakfast is served, the hostess greets me. No, thanks, I am still waiting for someone. After a moment, the person I am expecting arrives, and we get seated. I start the conversation by recalling our previous e-mail contact and my intentions when trying to set this meeting. He offers a couple of his latest books as a gift. Over the course of a full American breakfast, an invitation is made, and preliminary arrangements ensue for a research visit in the upcoming academic term.

### WAKING UP IN A POST-INDUSTRIAL WORLD

In the introduction to *Technology and the Lifeworld* (1990), Ihde performs this brief exercise where he recounts the first experiences of a typical day after waking up. The purpose of the exercise is to demonstrate the extent to which modern life is permeated by technological artifacts. Alarm clocks, beds, showerheads, clothes . . . indeed, as Ihde claims, "our existence is *technologically textured*" (1990, 1).

As a diligent student of Ihde, I attempt the same exercise. Yet, judging from the opening lines above, for a knowledge worker who wakes up in a networked, post-industrial society, existence is not only textured by mass-manufactured goods but by *services* too. And although technologies of the type identified by Ihde are also entangled in services relations, as they are above, they constitute only a part of these relations. The allusion here is to the forms of human-to-human exchange relying on touch, gesture, demeanor, and verbalization that are needed to get from bed to the breakfast table. Interpersonal contacts such as these occur in a variety of service settings, ranging from shopping, to public transportation, to physiotherapy sessions, and more.

Over the course of my PhD studies in the field of industrial design, postphenomenology has deeply informed my understanding of services (Secomandi, 2012). But its silence about the subject of interpersonal relations has also made me uneasy. The source of discomfort, to be sure, is not due to a complete disregard for the import of technology in service settings. As seen in some published studies, postphenomenological insights have been productively applied to service-related activities, including obstetric consultations (Verbeek, 2008a), microcredit projects (Selinger, 2009), diagnostics by clinical laboratories (Forss, 2012), and health improvement programs (Secomandi, 2013). In these studies, however, interpersonal relations receive scant attention and are at best treated secondarily to the role of material technologies. Selinger, perhaps, covers the most ground in the way of addressing the face-to-face dimension of services. His analysis of a microcredit project in Bangladesh takes note of how women who lend their mobile phones to the local population must remain silent and unobtrusive in the presence of customers, and how this contributes to shaping their identities as "phone ladies." But while Selinger (2009, 275) characterizes his study as postphenomenological because of a sensitivity for the ambiguity of human-technology relations, the analytical framework used to account for the constitution of "phone ladies" as such is mainly informed by Marxism and feminism (2009, 280–284). Currently, it remains unclear whether and how postphenomenology can contribute to our understanding of human-to-human service relations that require little, if any, intermediation of a technological artifact.

This chapter is intended as part of a program that seeks for fruitful resources within postphenomenology to deal with the matter of human-based service exchanges. Specifically, I want to know whether the body may be conceived as a kind of technological artifact that mediates human experiencing of the world, including relations among human beings. To achieve this goal, I will perform a close reading and critical examination of the decades-long pioneering contributions of Don Ihde in the field of the philosophy of technology. And while the work of other researchers who build upon or criticize his ideas is contemplated where appropriate, the analysis remains limited to published postphenomenological research.

The remainder of the chapter is organized as follows. In the section *Technique versus Technology*, I discuss what can be called an *eliminatory* approach to the human body qua technology, which is found in postphenomenological research. Particularly in Ihde's earlier philosophy of technology, the notion of *technique* is employed as referring to a type of embodied human activity that is somehow related to, yet lies beyond the scope of the "technological." In the interpretation that I propose, Ihde does not really convince us of any fundamental difference between technology and technique, but only reinforces their underlying similarities.

The following section, *Bodily Skill in Embodiment Relations*, expounds an approach that runs parallel to the previous one, but also encompasses more recent postphenomenological work. In this approach, a specific type of human technique—namely, *bodily skill*—is acknowledged as entangled in the learning of how to use a technology. I call this approach *subordinative*, because although Ihde and others eventually observe that bodily skills are interweaved in at least one type of human-technology relation, these skills are treated complementarily to the role of technologies and are not made significant on their own. In this section, I also explain how this restrained treatment of skills can be traced back to Ihde's dependence on three of his phenomenological predecessors: Heidegger, Husserl, and Merleau-Ponty.

The third and last section, *Experimental Phenomenology as a Self-Practice*, discusses a *prospective* approach that is gaining ground within postphenomenological studies. Here, skills are openly thematized as constitutive of human-technology relations and even treated as an object of conscious development and refinement by human beings. Special attention is given to Dorrestijn's and Verbeek's appropriation of the Foucauldian notion of *self-practice* as a way to understand the constitution of technologically mediated subjects. In concluding the inquiry, I return counterintuitively to Ihde's early introduction to phenomenology and present his variational method as one case of a self-practice for postphenomenologists. Thus, in answering the question "How can postphenomenologists deal with the matter of human technique," I anticipate—with some humor—that they can do so by attending thoroughly to the phenomenon and performing multiple variations of it.

## TECHNIQUE VERSUS TECHNOLOGY[1]

Whereas it is the case, as mentioned in the introduction, that an approach to the human body *as* technology can be born out of postphenomenology, it is also true that Ihde's seminal book in the philosophy of technology, *Technics and Praxis* (1979), initially gets us stranded. Or has he not dismissed in that book the technological nature of the body with the argument that "all that is important . . . is to note that *if* the body were an 'instrument,' it is indeed a very different one that those we use" (Ihde 1979, 80 n1)? That Ihde states this briefly in an endnote is uncharacteristic of his argumentative style. As others already observed (e.g., Sobchak 2006), he usually starts with some anecdotal evidence that covers familiar experiences with technologies, before adding progressive phenomenological sophistication to the analyses. In this case, instead of explaining—even if preliminarily—how the human body is unlike the "instruments" he chooses for a careful study, the difference is simply presumed and further ignored.

Later, in *Technology and the Lifeworld* (1990), the "instrumentality" of the human body will be mentioned again by Ihde through the notion of *technique*. The term, however, is framed in such a way as to mark it off from the domain of technology proper. In two exemplary cases, Ihde relies in his typically broad purview of historical and global developments in order to contrast contemporary Western, "technologically maximalist" cultures with the ancient, "technologically minimalist" cultures of the Australian Aboriginals (1990, 118–123) and the South Pacific islanders (1990, 146–150). According to Ihde, compared to other societies, the Aboriginals devoted extensive time to "leisure activities," which included storytelling, discussions about their mythological system, long communal festivities, art production, and so forth. Besides having invented few and simple technologies, including throwing weapons, basketry, and religious articles, the Aboriginals were nonetheless able to thrive in a harsh environment, an accomplishment that would be later admired by their Western colonizers. For this end, the Aboriginals relied on "technical knowledge" that Ihde qualifies as "vast and complex" (Ihde 1990, 121): sophisticated procedures to search for and prepare food, birth control methods involving body modification, and rule-governed hunting behavior to conserve the natural habitat from overexploitation.

Similarly, in the case of the South Pacific islanders, Ihde contrasts their navigational practices with modern, instrumentally pervasive modes of navigation. He observes that, except for the invention of sophisticated multihull boats, South Sea navigation relied primarily on observations of bird flying routes, star paths, cloud formations, wave patterns, as well as on the memorization of way-finding information through songs and on a peculiar sense of body positionality in space. Ihde concludes that the forms of navigation of both South Pacific islanders and Westerners were successful in finding new land across oceans, the attainments of the first only more remarkable for the tiny land masses they were able to discover.

At least one other critic was troubled by Ihde's characterization of South Pacific navigation as mainly non-technological, since the practice is evidently instrumental and goal-oriented (Hickman 2008). Ihde conceded to this point (2008a). But what is revealing about his comparisons is the choice not to contrast modern technological practices to patently less technologized traditions of the ancients, such as the Aboriginals' leisure activities. By emphasizing, instead, ancient practices that are functionally equivalent to modern, technology-intensive ones, as in the case of navigation, Aboriginals and South Pacific islanders are portrayed as no less instrumentalists than other contemporary societies. Presumably, their technological minimalism is warranted by the use of "instruments" closely associated with the human body, rather than material artifacts. What is important to note, however, is not that the recognition of similarities between technique and technology would undermine Ihde's demarcation between technologically maximalist and minimalist cultures. After all, he has also boldly stated that *"human activity from immemorial time and across the diversity of cultures has always been technologically embedded"* (1990, 20). The problem is that by foreclosing the topic of technique to a postphenomenological analysis, one is hindered from discovering the ways in which human experience of the world might be transformed solely through the application of "bodily instruments," independently of other forms of technologies.

There is a passage in *Technology and the Lifeworld* (1990, 26–27), which is the closest Ihde gets to acknowledging that techniques can be compared to technologies. It contains the following quote:

> There may be techniques with or without technologies. A sexual "technique" is not in itself a technology (although, in a derived and secondary sense, if such a technique is modeled after some mechanical process, there may be an interpretive connection between the two). Equally, techniques may be closely related to technologies particularly regarding patterns of use. (Ihde, 1990, 26)

Despite the hint to a possible "interpretive connection," the grotesque mechanical analogy steers readers away from wanting any prolonged explanation why techniques (sexual or not) are not themselves technologies. To my knowledge, Ihde does not elaborate any further than this on how a technique may be modeled after a mechanical process, or how the usage patterns of techniques relate to those of technologies. The remainder of the passage, nonetheless, presents clues to the basis of his differentiation. In it, Ihde offers preliminary clarifications about his notion of technology, before moving onto more detailed phenomenological descriptions in the book. He holds that the *materiality* of technologies correlates with the concreteness of our own bodies, and therefore must be retained. Moreover, technologies occupy an intermediating position as *artifacts* between "humans" and "world." In other words, technology constitutes a domain of experience that is not entirely encompassed by the human body, nor completely present as a worldly "other." If the latter were the case, Ihde argues, technology would be wrongly taken to be an object that can exist independently from concrete contexts of human practice. On the other hand, equating technology with the human would carry the connotation of "technique," or a totalizing form of "practice and thought" that glosses over the particularities of technological experience. Therefore, from the standpoint articulated by Ihde, what becomes intimately associated with the concept of technology is its *material artifactuality*. Techniques would not seem to possess that.

## BODILY SKILL IN EMBODIMENT RELATIONS

As seen in the previous section, Ihde initially rejects that human technique constitutes a genuine form of technology. In parallel to this, he also offers some detail about how a specific type of technique—namely, *bodily skill*—is entangled in human-technology relations, at least in relations of the type known as *embodiment*. Embodiment relations are occasions where technologies become extensions of the experiencing human body (Ihde, 1990, 72–80). An example is the use of eyeglasses. After being properly accommodated by the body, eyeglasses extend and transform innate visual capacities. Although Ihde does not connect his account of embodiment relations to the aforementioned interpretive connection between technique and technology, it is possible to gain some insight into the "derived and secondary sense" in which the first is said to be related to the latter.

In reality, bodily skills are rarely in the forefront of Ihde's detailed descriptions of the process of embodying various technologies. *Technics and Praxis* (1979), for instance, contains little or no references to them. This lack of attention might be due to Ihde's well-known inspiration in Heidegger's paradigmatic "hammer analysis." Only very recently has Ihde (2012a, 374) mentioned that Heidegger's analysis contains a blind spot, inasmuch as the embodiment of the hammer presupposes the previous attainment of handling skills by users. This realization now enables Ihde (2010, 120–127) to criticize Heidegger's negative appraisal of the typewriter, with the ironical observation that Heidegger possibly never became a skilled typist. Had the typewriter been properly embodied in the act of writing, Heidegger might have shown less a strong preference for fountain pens.

Despite the Heideggerian influence, from *Technology and the Lifeworld* (1990) on Ihde starts to contemplate the topic of bodily skills as part of his descriptions of human-technology relations of embodiment. At times, he will point to the body as a kind of "fringe" phenomenon one needs to "adjust to" in the process of learning to use a technology. About the use of eyeglasses, Ihde writes:

> The very first time I put on my glasses, I see the now-corrected world. The adjustments I have to make are not usually focal irritations but fringe ones

> (such as the adjustment to backglare and the slight changes in spatial motility). But once learned, the embodiment relation can be more precisely described as one in which the technology becomes maximally "transparent." It is, as it were, taken into my own perceptual-bodily self experience. (Ihde, 1990, 73)

With the exception of a recent detailed case on the embodiment of hearing aids (Ihde, 2007, 243–250), rarely do bodily adjustments to technologies receive greater attention than above. The use of handheld telescopes, in *Expanding Hermeneutics* (1998, 153–157), is one such case that deserves special attention. Ihde argues that when an object like the Moon is observed through the telescope, it is also transformed by it; what is experienced is displaced from its original context, magnified, and seen to contain mountains, craters, and so on. These transformations caused by the telescope are accompanied by a simultaneous alteration in bodily position, in the sense that one experiences being "closer" to the Moon. Additionally, if the telescope user is a beginner, a certain wavering of the object will be noticed reflecting the micromotions of his own body. In order to "fix" the phenomenon observed through the telescope, beginners must be able to stabilize their bodies accordingly. Just like in the case of eyeglasses, the point is that the body becomes noticeable in the process of attaining the kind of maximally transparent experiences that characterize human-technology relations of embodiment.

Yet, two suspicions can be raised about how the human body appears in Ihde's account of embodiment relations. First, the body surfaces in the context of the experience in a *negative* way, in situations where the technology is not truly embodied by the user. Ihde argues that when someone engages incorporable technologies, like eyeglasses and telescopes, the body becomes detected as an artifact that partially obfuscates the desired transparent relation to what is perceived. Hence, novice users realize that their trembling holding of the telescope disturbs the stable display of the Moon. Then, in the process of becoming a skilled user, at the same time that the technology becomes an extension of perceptual capacities, the bodily intrusion seems to disappear from the experience. In other words, the artifactual body of the beginner must vanish for the embodiment relation of the expert to obtain.

What I am doing here is to apply the same formula concocted by Ihde (1979, 103–129) to remedy a certain negativity in Heidegger's tool analysis. Although for Heidegger the experience of technologies in an embodied condition preceded and founded any knowledge one can have *of* them, he employed an indirect tactic to demonstrate this, where the technological artifact was made explicit in situations of "breakdown" (when it breaks, malfunctions, or goes missing). According to Ihde,

> In spite of the phenomenological correctness of Heidegger's analysis, the *negative* way in which the instrument emerges from transparency in use in his

> analysis casts a sense of disvalue upon any positive thematization of an instrument. (1979, 28)

I believe that a similar negativity ensues as the unskilled body is revealed in situations where embodiment human-technology relations break down. Thus, the trembling hands of the telescope user and the altered spatial motility of eyeglasses wearers are made salient as obstructions for the proper embodiment of technologies, not as positive features of the process of developing a bodily skill.

The second suspicion about Ihde's analysis of embodiment relations is in regard to how the human body is recovered *indirectly* from what is focal in the experience. The novice user experiences *immediately* the trembling Moon, and *reflexively* the wavering hand: "The Moon's magnified character is simultaneously the *reflexive magnification of my bodily motion*" (Ihde 1998, 155). Ihde structures these events in terms of a *noema-noesis* correlation, what points to a terminology that was more fully explained in the earlier book *Experimental Phenomenology: An Introduction* (1986, 29–54), first published in 1977. In that book, the noema and the noesis are introduced as parts of the phenomenological notion of *intentionality*, which represents the primordial connectedness between humans and world that grounds all possible knowledge. Following Husserl, Ihde explains that the noema represents the end pole for what is experienced, while the noesis represents the mode of experiencing that which is experienced. He further clarifies the order of the correlation, stating that the noetic domain cannot be known directly, but only reflexively after examination of the noematic one.

The first edition of *Experimental Phenomenology* is not in its core devoted to technologies. Nonetheless, the claim for the priority of the noema is initially supported with a brief example of a human-technology relation of embodiment. When chopping wood with an ax, Ihde notes, "my primary energy and concentration is focused *almost* totally in the project itself. . . . 'I' should be put roughly as 'I-am-in-the-ax-directed-towards-the-wood'" (Ihde 1986, 47). In other words, the "I" as the bearer of the experience is mostly implicit in the technological activity. This is not to say that any sense of "I" is definitively ignored. But, phenomenologically, the "I" is mostly situated in the noetic domain, and therefore to be recovered after carefully attending to the noema: "The 'I,' particularly in its thematized form, comes late in the analysis rather than being given as a first" (Ihde, 1986, 48).

By making the "I" mostly unknown, at least initially in the analysis, the woodchopper as a subject in the activity is in effect "black-boxed." Ihde is well aware of this move, and he explains that it strategically protects the phenomenological method from becoming a naïve form of introspection:

> An introspective ego or "I" claims direct, immediate and full-blown self-awareness as an initial and given certain. In phenomenology, the "I" appears by means of and through reflection upon the phenomena that in toto are the world. (Ihde, 1986, 50–51)

However, if by "I" Ihde means to embrace the woodchopper as physical being, the tendency is to automatically submerge his handling skills into the noetic domain too. This tendency is observed again as Ihde proceeds with more thorough analyses of visual phenomena, later, in the book. In the next section, I will comment on that part in order to highlight what I see as fruitful outcomes of Ihde's methodology. For the moment, the point is that the location of the body within the noetic domain of experience could be preventing postphenomenology from foregrounding the role of bodily skills in human-technology relations.

Would it be possible, then, to conceive of aspects of the human body as noema? An answer to this question might be found in the growing stream of postphenomenological studies that expands on our understanding of embodiment relations by examining the cases of implants and prostheses (Verbeek, 2008b; Welton, 2006; De Preester, 2011a; 2011b; Lenay, 2012; Ihde, 2008b, 31–42; 2012a). Because of the philosophical issues that are raised when material artifacts get more deeply or permanently attached to the human flesh, this research sheds interesting light onto the blurred boundaries between bodies and technologies.

Of special importance among the references listed above is Welton (2006), because of his clarifications over phenomenological notions of the human body. Welton wants to push Ihde's notion of embodiment relations towards more "cyborgian" forms of integration, for example, when a brain implant allows someone to regain eyesight. I will not summarize the whole of his contribution, but only note that it involves a critical comparison between accounts of the body by Husserl and Merleau-Ponty. Welton ends up favoring the Merleau-Pontean account, and drawing on refinements advanced in neuropsychological research, he proposes two distinguishable, yet complexly entangled dimensions of experiencing. On the one hand, there is the *body schema*, which according to Welton,

> is not an object expressly known but only "tacitly understood." It functions not as object of consciousness because it forms the basis of or the hidden "background" behind all explicit intentional acts. . . . The awareness that attends the body [schema] as it engages the world . . . is not a perception of the body. (2006, 201)

And there are the *body images*, which in turn,

> arise from an explicit perception of the body or, more generally, from our conceptions of the body and our feelings about our body. They are representations that arise when my body is taken as an object, for example when I look at my body in a mirror or gaze at the body of others, or when I describe the body that I see in the language of my culture, or when I assume a certain emotional attitude toward my body. (Welton, 2006, 201)

Welton, then, superimposes these two dimensions onto the human-technology relation of embodiment and adds a twist. He argues that when someone learns to hit a ball with a baseball bat, the embodiment process involves body images insofar as one starts by "visualizing" how the bat relates to the concrete body, as well as the correct ways of holding it (Welton, 2006, 203). After reaching the goal of technological embodiment, the body schema gets transformed and materially extended by the artifact, thereby making one capable of experiencing the hit of the ball at the end of the bat.

Based on this two-pronged understanding of the body, Welton describes the process of technological embodiment in a slightly different way than Ihde. When first encountering an incorporable technology, an aspect of the unskilled user's body "reaches out" to the artifact to be embodied, rather than "intrudes" upon an already established relation. Hidden in Welton's analysis is a positive evaluation of how body images stir the development of skills. Nevertheless, similarly to Ihde's account, once extended so to subsume the technological artifact, the body schema of skilled users does away with the artifactual body of the novice. Apparently, for Welton, usage skills are not experienced as body image phenomena in consolidated human-technology relations of embodiment.

In his response to Welton, Ihde (2006, 284–285) sides with the critique of Husserl and positions himself also closer to Merleau-Ponty. To understand Ihde's positioning we must recall that he, too, has characterized the human body as two-fold. In the book *Bodies in Technologies* (2002), Ihde differentiates between the *here-body* and the *image-body*; the here-body being defined as the "quasi-primary," "fully sensory" body, and the image-body as the "partially disembodied," "quasi-other" body (Ihde, 2002, 3–15). Ihde conceives of a dialectic existing between these two dimensions that is, in principle, similar to Welton's explanation of Merleau-Ponty's views. However, the quote below contains an ambiguity that could undermine the affinity:

> It is the here-body in action that provides the centered norm of myself-as-body . . . it should also be noted that such a body experience is one that is not simply coextensive with a body outline or one's skin. The intentionality of bodily action goes beyond one's bodily limits—but only within a regional, limited range. A good example may be taken from martial arts experience wherein one can "feel" the aimed blows even from behind and aims one's own

> activity beyond any simple now-point. One's "skin" is at best polymorphically ambiguous, and, even without material extension, the sense of the here-body exceeds its physical bounds. (Ihde, 2002, 6)

The passage is particularly relevant because, differently from the cases discussed so far, Ihde mentions the body's ability to "extend" its experiential reach *without* the use of mediating technologies. The same example was presented before in a slightly different format in *Technology and the Lifeworld* (Ihde, 1990, 74). In that book, the example was part of a discussion about embodiment relations, suggesting that martial arts skills might somehow be embodied like other technological artifacts. In *Bodies in Technologies*, the example does not suggest that skills pertain to the image-body, only accentuates the malleability of the here-body.

But if Welton's disambiguation between the two phenomenological senses of the body is correct, the passage above becomes problematic. This is because Ihde first writes that "one's own activity"—say, karate blows—may be extended beyond the outline of the skin *as here-body*. Then, he states that one can "aim" these blows, what would appear to require having some experience *of* them. Using Welton's vocabulary, the blows integrate the body schema of karate practitioners at the same time that they would be explicitly experienced as body image. This interpretation is supported by the following claim:

> One can simultaneously experience one's here-body *from its inner core* while having a partial, but only partial, "external" perception. (Ihde, 2002, 6; emphasis added)

But if I experience my *here-body* "from within," would that not make it a partially disembodied body, an *image-body*, to stick to Ihde's terms? Are not the blows of martial arts practitioners more accurately defined as image-body artifacts that mediate and transform their experiences of the world and the self?

As an interim conclusion to the argumentation developed thus far, it seems possible to draw on an existing phenomenological discrimination between two senses of the human body and associate skills with the notion of the image-body (or body image, following Welton). Because in *Experimental Phenomenology* Ihde does not rely on this distinction, both the image-body and the here-body are conflated into the noetic domain of human experience. The elimination of the human body from the noematic domain carries over to Ihde's philosophy of technology of *Technics and Praxis* and subsequent works, where "technology" becomes roughly equated with the noema and "human" with the noesis. As a result, skills—or, more generally, techniques—could never be genuinely approached as bodily artifacts in human-world relations. At best, they had to be derived indirectly, often negatively, from human practices involving other sorts of technologies.

## EXPERIMENTAL PHENOMENOLOGY AS A SELF-PRACTICE

A developing strand of postphenomenological work now starts to conceive of bodily skills as constitutive, not merely appended to, our relations to technologies (Ihde, 2009, 16–19; 2010, 128–139; 2012b, 129–184; Rosenberger, 2009; 2013; Verbeek, 2011, 66–89; Dorrestijn, 2012). I start with some of Ihde's recent publications, including the important second edition of *Experimental Phenomenology* (2012b).

As mentioned in the previous section, the first edition of *Experimental Phenomenology* does not take on systematic analyses of technologies. The book is mainly dedicated to a hands-on introduction to the practice of phenomenology using simple line drawings of visual illusions as experimental material (Ihde, 1986, 67–121). Although Ihde praises Husserl for inventing the so-called variational method to arrive at the *essence* of a phenomenon, his own rigorous application of this method yielded something different—the *multistability* of these visual illusions. The originality of this finding was at first not fully grasped by Ihde (2012b, xi, xiv). Still, the notion of multistability was transposed by him to the domain of technology, leading to claims regarding the multistable nature of such practices as oceanic navigation (Ihde, 1990, 146–150) and writing (Ihde, 2010, 128–139). In technological contexts, multistability means that a "same" technology can have multiple instantiations in history or across cultures, which are coherent to the particular circumstances of use.

The second edition of *Experimental Phenomenology*, now subtitled *Multistabilities*, is expanded with case studies on the multistability of audiovisual and communication technologies (mainly screen-based) (Ihde, 2012b, 131–143, 145–153), the camera obscura (2012b, 155–169), and bow-under-tension technologies, of which archery is the primary example (2012b, 171–184, see also Ihde, 2009, 16–19). The cases on writing devices and archery, in particular, contain unusual elaboration over the topic of bodily skills. Only archery is considered here, even though Ihde's analysis of writing devices bears similar insights.

It is not necessary to repeat all the different types of archery that Ihde discusses, but simply recall the usage skills pertaining to two. Thus, in the case of the English Longbow, the person stands up holding the bow with one arm stretched to the front, and then pulls back the string while holding the arrow with all of his fingers. And, in the case of the Mongolian Horsebow, the bowman, now mounted on a galloping horse, draws the bowstring toward his cheek as he simultaneously pushes the bow away from the body, also trying to synchronize the arrow's release with the moment when the horse's hoofs come off the ground.

Although differences in bow size, material, reach, and other factors are also mentioned in text, when compared to the descriptions of other technologies, the portrayal of archery stands out for the prominent role granted to bodily skills in defining the multistabilities of the technology. Unfortunately, because in the case of technological multistability Ihde does not explain the procedures of his variational method as carefully as in the case of visual illusions, it is unclear how skills would be framed following the noema-noesis correlation that was introduced before.

Still on the topic of technological multistability, Rosenberger (2009, 176) has coined the term *relational strategies* to refer to "the particular configuration of bodily habits, intentions, and conceptions that make it possible for a person to take up a particular stable relation [to a technology]." Taking web navigation as an example, Rosenberger argues that when a person who surfs the web comes to a slowly loading page, a shift of relational strategy typically occurs, accompanied by a sudden drop in transparency. In line with the breakdown phenomenon described before, the drop makes the user aware of her embodied interaction with the keyboard, mouse, screen icons, and so on. Alternatively, users may also adopt a different relational strategy towards slowly loading webpages, implicating a different set of bodily comportments and habits that are more adequate for the situation. Thus, the next time a sluggish webpage appears, a user can shift to this other relational strategy without feeling a significant drop in transparency (she might, for instance, revert to working offline on an already open document).

Although Rosenberger is not specific about this issue, for the drop in transparency to be annulled, the shift in relational strategy would presumably happen automatically, without the user consciously deciding which bodily configuration to adopt in respect to the technology. What Rosenberger brings to the current discussion is the possibility to anticipate and deliberately develop techniques of use. In another paper, he argues that there are *abstract* relational strategies that can be generally applied across different use scenarios or with technologies alike (Rosenberger, 2013).

The idea that techniques can be an object of planning is taken up a notch by Dorrestijn (2012) and Verbeek (2011). These researchers are interested in clarifying what sort of *human subjects* emerge from technologically mediated existence. For that end, Dorrestijn creates a bridge between postphenomenology's interest in technology and Foucault's work on subject constitution, especially the later contributions to ethical theory. Dorrestijn argues that there are certain routines involved in our relations to technology that represent a form of "disciplining" of the body, and that these do not just happen or get imposed upon humans by some external reality; humans actively co-shape their own technologically mediated subjectivity. One way in which they can do this is by rehearsing what Foucault called "technologies, or practices of the self" (Dorrestijn, 2012, 236). It is still unclear how the concept of self-practice relates to phenomenology's framework of intentionality. Nonetheless, for Dorrestijn, in technological contexts self-practices are

> activities whereby people get attached to technologies and accommodate mediation effects in their existence. . . . For example, training of technically mediated gestures is necessary even to be able to sleep in a bed. It may seem convenient, comfortable, and very natural, but it has to be trained. Children frequently drop out of their beds. In addition, they need training and habituation to stop them wetting their beds. Once these skills have become routines, one tends to forget ever having learned them. (2012, 236)

Verbeek (2011) also adopts the Foucauldian notion of self-practice in his investigation into the morality of technological artifacts. In respect to ultrasound devices, he argues that this technology partly determines how expecting parents are constituted as moral subjects. On the other hand, parents themselves must critically engage practices of ultrasound screening and negotiate the sort of moral subjects they wish to become. For instance, they can make use of ultrasound during prenatal examinations only to determine an expected date of birth, or choose to do all diagnostic tests possible, or even refuse to use the technology. Verbeek (2011, 89) goes as far as to suggest that self-practices can be purposively adopted for "styling" or "designing" one's own technologically mediated subjectivity.

To reinforce this recent attunement of postphenomenological research to the issue of human technique, I finalize by returning to Ihde's *Experimental Phenomenology* and offering a somewhat idiosyncratic reading of that book. In the last section, I stated that the book's core about the interpretation of visual phenomena changed little in the way that bodily skills were blackboxed by the variational method. Now I argue that it is precisely in that part that one finds the richest case on the nurturing of a skill, in this case, a skill for seeing.

On a first level of interpretation, what Ihde does in *Experimental Phenomenology* is to provide instructions that allow readers to confront deeply sedimented seeing habits in order to develop new abilities for uncovering the multistability of visual phenomena. I hold that underlying Ihde's text is a bodily skill being transmitted that functions like a self-practice for aspiring postphenomenologists. Selinger (2006, 92), before, noted a similar quality regarding the use of the variational method. By revealing the many possible interpretations of phenomena, the technique allows Ihde to question extant perceptual prejudices and, as such, it is constitutitive of his own subjectivity.

Acknowledging the sort of bodily skills that are implicated in the variational method, however, requires not simply reading Ihde but questioning:

How is one able to experiment with the phenomenological practice that he wants to pass on? Then, one starts to notice several references, at times subtle ones, to particular bodily comportments that must be assumed and perfected by readers to apply the method successfully. I turn to one poignant example of this: the "curved line" drawing, found in chapter 5 (Ihde, 1986, 81–90). In this chapter, Ihde is trying to show that what is ordinarily experienced as curved lines in the drawing, under phenomenological scrutiny can be experienced as straight lines. To facilitate this perception by readers, he gives these instructions:

> Look at the drawing in the following way: First, *focus* your gaze intently upon the vertex, where all the diagonal lines converge in the center of the figure. Second, deliberately see the vertex as three dimensional, and in the far distant background, that is, push the vertex back, as it were, until the diagonal lines are seen to lead to infinity. Granting that this takes a certain amount of concentration, subjects usually can do the task quickly, and then the formerly curved horizontal lines appear straight. But this is so only as long as the subject focuses upon the vertex intensely, making the horizontal lines peripheral to the central focus. (Ihde, 1986, 84)

I get better accommodated in the chair and bring the book upwards, closer to my face and more in line with my head level. I am drawn to the figure that now faces me. At the same time, I also notice the weight of the book on my fingertips and hands, the pressure points where my elbows touch the chair's armrests, the erect posture of my neck and back, and even a slight dizziness caused by the proximity of the figure upon which my gaze is focused. Addressing this explicitation of my body as causing a drop in transparency, however, would be misleading. Because the straight lines have not yet *fulfilled* in experience (i.e., they have not yet been *seen* by me), it is inaccurate to explain the situation as one of breakdown. If there remains a peculiar feeling of inadequacy and obtrusiveness of my bodily parts as something that requires immediate attention before I can concentrate on the figure, from a phenomenological standpoint, this is a positive feature of me trying to follow the instructions provided in text. Granted, once I get to see the lines as straight, these bodily artifacts fall to the background of perception, only to protrude again if I fail to repeat the exercise or if I need to adjust the learned bodily comportment in another visual experiment. But, even as my body becomes tacit, is the newly acquired skill irrecoverable from the experiential act? Do I not deliberately utilize my body in this or that manner when I want to switch between seeing curved or straight lines?

What I am suggesting with this admittedly brief analysis is that it would be difficult to learn the variational method from Ihde without relating to aspects of my body as quasi-others, or image-bodies. As these image-body phenomena get crafted in the form of skills, then they are partially integrated as extensions of my here-body. This integration is not much unlike what happens when telescopes, eyeglasses, and baseball bats enter human-technology relations of embodiment. Although skills are certainly less dissociable from the human body, their artifactual condition may still be positively noted. In fact, the thrust of any postphenomenological analysis rests precisely on the possibility of discerning the material influence of artifacts in human-world relations. This is what prevents postphenomenologists from slipping into dystopian beliefs about the totalizing effect of technology on humanity. This is also what clears the ground for the emergence of a rich vocabulary speaking of modes of human-technology-world relations (embodiment, hermeneutic, etc.), amplification/reduction structures, technological trajectories, and so forth. Similarly to other mediating technologies, techniques can transform human experiences without offering a single sweeping interpretation of the world and of other human beings. The variational method would seem to prove the exact contrary of this. When rigorously applied, the technique yields the multiplicitous appearances of phenomena. In the process, as self-practice, it takes part in the becoming of a postphenomenologist.

## ACKNOWLEDGMENTS

The beginning of this chapter recounts the events leading to my first face-to-face meeting with Don Ihde after so many encounters intermediated by books. I am most grateful for his invitation to come to Stony Book as a visiting researcher. The thoughts expressed in this chapter were first presented there, in March 2012, to the participants of Ihde's *Technoscience Seminar*. Parts of the text were later included in my PhD thesis at the TU Delft, under the supervision of Petra Badke-Schaub and Dirk Snelders. I want to thank all of the people mentioned, plus the editors of this collection, for their encouragement.

## NOTE

1. This section and the following one contain parts of chapter 5 of my PhD thesis (Secomandi, 2012).

## REFERENCES

De Preester, H. (2011a). "Technology and the Body: The (Im)Possibilities of Re-Embodiment." *Foundations of Science* 16 (2–3): 119–137.

De Preester, H. (2011b). "Technology and the Myth of 'Natural Man.'" *Foundations of Science* 17 (4): 385–390.

Dorrestijn, S. (2012). "Technical Mediation and Subjectivation: Tracing and Extending Foucault's Philosophy of Technology." *Philosophy & Technology* 25 (2): 221–241.

Forss, A. (2012). "Cells and the (Imaginary) Patient: The Multistable Practitioner–Technology–Cell Interface in the Cytology Laboratory." *Medicine, Health Care and Philosophy* 15: 295–308.

Hickman, L. (2008). "Postphenomenology and Pragmatism: Closer Than You Might Think?" *Techné* 12 (2): 99–104.

Ihde, D. (1979). *Technics and Praxis*, vol. XXIV, Boston Studies in the Philosophy of Science. Boston, MA: Reidel.

Ihde, D. (1986). *Experimental Phenomenology: An Introduction.* Albany: State University of New York Press.

Ihde, D. (1990). *Technology and the Lifeworld: From Garden to Earth.* Bloomington: Indiana University Press.

Ihde, D. (1998). *Expanding Hermeneutics: Visualism in Science*. Evanston, IL: Northwestern University Press.

Ihde, D. (2002). *Bodies in Technology*. Minneapolis: University of Minnesota Press.

Ihde, D. (2006). "Forty Years in the Wilderness." In E. Selinger (ed.), *Postphenomenology: A Critical Companion to Ihde*. Albany: State University of New York Press, 267–290.

Ihde, D. (2007). *Listening and Voice: Phenomenologies of Sound.* 2nd ed. New York: State University of New York Press.

Ihde, D. (2008a). "The Corpus Is Not Yet Closed. . . ." *Techné* 12 (2): 126–132.

Ihde, D. (2008b). *Ironic Technics*. New York: Automatic Press/VIP.

Ihde, D. (2009). *Postphenomenology and Technoscience: The Peking University Lectures.* Albany: State University of New York Press.

Ihde, D. (2010). *Heidegger's Technologies: Postphenomenological Perspectives*. Perspectives in Continental Philosophy. New York: Fordham University Press.

Ihde, D. (2012a). "Postphenomenological Re-Embodiment." *Foundations of Science* 17 (4): 373–377.

Ihde, D. (2012b). *Experimental Phenomenology: Multistabilities*. Second Edition. Albany: State University of New York Press.

Lenay, C. (2012). "Separability and Technical Constitution." *Foundations of Science* 17 (4): 379–384.

Rosenberger, R. (2009). "The Sudden Experience of the Computer." *AI & Society* 24 (2): 173–180.

Rosenberger, R. (2013). "The Importance of Generalized Bodily Habits for a Future World of Ubiquitous Computing." *AI & Society* 28 (3): 289–296.

Secomandi, F. (2012). "Interface Matters: Postphenomenological Perspectives on Service Design." Doctoral Thesis, Delft: Delft University of Technology.

Secomandi, F. (2013). "Thinking through the Service Interface: A Study of Philips DirectLife." *Design Philosophy Papers*, no. 1 / 2013.

Selinger, E. (2006). "Normative Phenomenology: Reflections on Ihde's Significant Nudging." In E. Selinger (ed.), *Postphenomenology: A Critical Companion to Ihde*. New York: State University of New York Press, 89–107.

Selinger, E. (2009). "Technology Transfer and Globalization: A New Wave for Philosophy of Technology." In J. K. B. Olsen, E. Selinger, and S. Riis (eds.), *New Waves in Philosophy of Technology*. New York: Palgrave MacMillan.

Sobchak, V. (2006). "Simple Grounds: At Home in Experience." In E. Selinger (ed.), *Postphenomenology: A Critical Companion to Ihde*. New York: State University of New York Press.

Verbeek, P-P. (2008a). "Obstetric Ultrasound and the Technological Mediation of Morality: A Postphenomenological Analysis." *Human Studies* 31 (1): 11–26.

Verbeek, P-P. (2008b). "Cyborg Intentionality: Rethinking the Phenomenology of Human–Technology Relations." *Phenomenology and the Cognitive Sciences* 7 (3): 387–395.

Verbeek, P-P. (2011). *Moralizing Technology: Understanding and Designing the Morality of Things*. Chicago: University of Chicago Press.

Welton, D. (2006). "Body and Machines." In E. Seligner (ed.), *Postphenomenology: A Critical Companion to Ihde*. Albany: State University of New York Press, 197–209.

## *Chapter Seven*

# Four Dimensions of Technological Mediation

Asle H. Kiran

### THE TWO-SIDEDNESS OF TECHNOLOGICAL MEDIATION

At the center of postphenomenological thinking is *technological mediation*. Don Ihde's typology of human-world relations—embodiment, hermeneutic, alterity, and background—provides a methodologically suggestive repertoire of concepts dealing with how technologies shape and reshape both "humans" and "world" through *specific* mediating effects (Ihde, 1979; 1990). However, since shaping means to accentuate certain characteristics of worldly phenomena (or in humans)—what Ihde (1990, 76) calls *magnification*, there can be no shaping movement without a corresponding downplaying movement—what he calls *reduction*. As an example, consider the transformation of the moon that happened when Galileo Galilei first viewed it through his telescope; in order to become a better scientific object "the Moon became larger, magnified. But was also displaced—telescopically it was taken out of the night sky and relocated within the field of telescopic vision. It lost its place in the expanse of heavens and became a more focal, particularized object" (Ihde, 2002, 58). The accentuation of the moon as a scientific object meant that the moon as a lifeworld object moved into the background. At least for the time the moon was put under scientific scrutiny, that is, as long as this particular technological mediation was active.

The *magnification-reduction structure* exemplifies a two-sidedness that we find in all technological mediations. However, rather than taking magnification as a "positive" side and reduction as a "negative" side to our ways of dealing with technologies, the two-sidedness should be understood as demonstrating a very fundamental manner in which we *are* in our technological lifeworld. As such, far from being a mere empirical fact about technological mediation, the two-sidedness instead expresses an existentially significant constraint in how technological mediations shape both "world" and "human." Furthermore, the two-sidedness also has methodological consequences when it comes to technology assessment; very often we will find that "the question of" an emerging technology is not whether it is "good" or "bad," or if we can choose whether it should enter our lives or not, but *how* we can shape our lives in relationship *with* this technology. Exposing the two-sidedness, therefore, contributes to an important broadening of the basis on which we assess technologies, both on a societal level and on a personal one.

In this chapter, I shall explore different sides to the technological shaping of our lifeworld by showing how the two-sidedness is expressed in four different *dimensions* of technological mediation. The concept of "dimension" conveys that technological mediation can, and should, be analyzed from different perspectives. In their existential dimension, technologies involve a *revealing-concealing* structure, constituting what kind of world we find ourselves in. In their epistemological dimension, technologies are *magnification-reduction* structures, thereby shaping the type of knowledge we have about the world. Through the practical dimension, technological mediations exhibit an *enabling-constraining* structure that shapes action and behavior. And finally, I shall also look at how the ethical dimension of the two-sidedness expresses an *involving-alienating* structure, in which technologies both support and deplete our efforts to chisel out a *good life* for ourselves.

As I said, one purpose of focusing on the two-sidedness of technological mediation is its methodological implications. First, the two-sidedness displays how questions of quite different type—existential, ontological, epistemological, ethical, political, practical, methodological, and so on—can be connected within one perspective (the postphenomenological). Furthermore, the two-sidedness substantiates why technological mediation is an indispensable methodological nexus when we investigate, anticipate, assess, and attempt to influence the technological shaping of our society; although the dual concepts do not express positive and negative sides to technological mediation, all dimensions invite us to evaluate the difference between what is being brought to the front of our attention, and what is being pushed to the background of it. Also, bringing out the two-sidedness of technological mediations is helpful not just for technology assessment, but also for anthropological, sociological, and historical studies of technological cultures. Finally, and not merely because there is an ontological dimension; existentialist philosophical and psychological analyzes of our embeddedness will gain from this breakdown of technological mediation.

Although the two-sidedness in itself places no values on specific technologies, its consequences might be value-laden. As such, the consequences of technological mediation in all four dimensions carry normative aspects. However, the ethical dimension discloses that some specific technologies also manifest some explicit normative connotations. For instance, technologies that invade our privacy will for many be regarded as a negative technology. However, things are not straightforwardly negative and positive here either, as a negative pull from one type of technology might follow closely on some very positive aspects, such as the classic dilemma between being able to feel safer but also as being monitored by CCTV-cameras in public places. The ethical challenge then becomes to balance these two aspects of technology development, implementation and use. This is a good example of how the two-sidedness brings out ambivalence in all four dimensions of technological mediation; there are no "good" technologies and there are no "bad" technologies, not in relation to ethical issues, and not in relation to ontological, epistemological of practical issues either, but technologies certainly shape how we deal with such issues.

## THE ONTOLOGICAL DIMENSION OF TECHNOLOGICAL MEDIATION: REVEALING-CONCEALING

I begin with the most general and abstract dimension of technological mediation, namely how technologies shape our ontology. Or, to put it in different terms; how technologies shape the world in which we find ourselves. This might be taken to be a mere empirical statement; how the nuclear bomb shaped the post-war era, or how internet and mobile technologies have changed how we communicate. However, as having ontological impact, technological mediation shapes the world in a sense that is also a matter about shaping *us*, humans, as individuals and as societies. As such, this shaping has impact on how we perceive and act in the world, and how we see ourselves as being in that world.

One reason that I start with this dimension is because—from a philosophical point of view—it is the most fundamental and foundational and therefore the most important dimension for how we understands the other dimensions. But another, equally important reason for me to start here is that this dimension clearly displays that the two-sidedness permeating all dimensions of technological mediation should not be interpreted as a dilemma or a paradox; the dual concepts do not express a positive-negative relation. Instead, the two-sidedness has a certain inevitability to it; the manners in which we deal with the world is mediated in many ways—through symbols, language, culture and history, embodiment, and technology, and all these forms of mediation make something stand out and come into focus, while other things disappear or fade from view. The latter "movement" is necessary for the first movement to take place. In many ways, then, the dualities found in the different dimensions of technological mediation express a key dimension to our being-in-the-world.

In relating technology to ontology, it is impossible to bypass Martin Heidegger, who provided two different ontological arguments for how technologies contribute to shape the world into *a* world: the well-known *Zeug*-analysis from *Being and Time* (1927/1962), and the equally famous but less cherished analysis of modern technology from the essay "The Question Concerning Technology" (1954/1977). Although very different in tone, both analyses express tools and technologies as having a *revealing-concealing structure*. In the first analysis, Heidegger claims that when we are in-the-world, we do not find ourselves in a geometrically defined room, or space, surrounded by "neutral" or pure entities, instead, the space we are in is defined through the entities that fill it (1962, 95ff). However, this should not be understood to mean that the room is merely "the sum total" of these entities, because we get to know, firstly, the entities through a certain in-use-approach that defines them as the kind of entities that they are, and secondly, the space as connected to the entities in the *way they are*. The way we are able to approach an entity is attributable to its readiness-to-hand, its *zuhandenheit*, which defines how a material entity can be put to use. Readiness-to-hand is how the entity exists, how it *can be* approached by us. According to Heidegger, all entities are primarily approached as tools, or equipment as Heidegger calls them, as being useful for something. Equipment "includes everything we make use of domestically or in public life. In this broad ontological sense bridges, streets, street lamps are also items of equipment" (Heidegger, 1988, 292).

Through its readiness-to-hand, the equipment displays a structure of *in-order-to* that constitutes the relation between the particular entity and its environment—this is how equipment *reveals*. The structure of in-order-to involves a *referring* that is not a mere index; instead, it has an ontological character. That is to say, the referring serves a dual purpose: It reveals what serves as context for it, and through doing so, the entity in turn is constituted as a tool within the context because of its "specific thingly character" (Heidegger, 1962, 97f). Because of the way we approach entities and the relation between entities and their context, *we* are connected to the world in manners that are specific to the various technologies we deal with. What constitutes the world for us, then, is revealed through our engagement with equipment; and whatever does not belong to the context of this engagement is concealed. At least for this specific technological mediation.

Note that "context" engages not only material entities. In some, perhaps most, cases, the context will comprise non-material components as well. For cars, for instance, to "function technologically" legislations in various forms have been created, social institutions like the DMV have been established to support some of this legislation, a system of auto clubs like the Norwegian NAF, mandatory car insurance and so on, are also consequences of the car having its technological function within a society. Such non-material features are also essential parts of the world that is revealed to us. For some technologies at least, then, to grasp their revealing aspect, we need to look beyond the mediations we perform with them, and assess the broader organizational change they produce in a society. The revealing that happens through using social media is not confined to the actions of sitting in front of an online computer, to type and to read and so on, it is also related to the changing forms of communication, both in methods and content, which again has an effect on social relations, and so on.

The *Zeug*-analysis, which for the most part deals with "old" technologies, hammers, nails, and so on, does not invite normative judgment on the revealing-concealing structure. But in the later essay Heidegger is unambiguous about the specific revealing that happens through modern technology—primarily post-industrial technologies—as being a negative one. Here, Heidegger argues that modern technology has a specific essence, which he calls *Gestell*. This essence, he says, "is by no means anything technological" (1977, 4). Consequently, we will never disclose the essence of technology by scrutinizing technological items. *Gestell* is perhaps best understood as our attitude to the world that we assume when acting in the world using modern technologies. As such, *Gestell* manifests itself in humans through *modern* technological mediations. In putting modern technology to use, whatever it is that the technology operates on is "ordered" into being resources, *Bestand*, for the technological activity. For the sake of simplicity, we can think of this ordering as how we treat our surroundings due to the effect of modern technologies. In that regard, *Bestand* can be understood in two ways, one literally, where things in the world are lined up as resources, and then more figuratively, where it alludes to a state of mind, an *attitude* to the world that we cannot but take up as a result of how modern technology shapes our technological mediations.

The latter reading relates *Gestell* to a particular kind of thinking that Heidegger elsewhere refers to as *calculative thinking*. This concept refers to our tendency to be unscrupulously goal-oriented in our activities rather than contemplative on what it means to be *underway* to something. Calculative thinking "computes ever new, ever more promising and at the same time more economical possibilities. Calculative thinking races from one prospect to the next. Calculative thinking never stops, never collects itself" (Heidegger, 1969, 46; cf. 1977, 15). Both calculative thinking and *Gestell* push towards a maximizing of efficiency, ultimately not because it is for the best for humans and society, but because efficiency becomes a goal in itself.

For Heidegger, modern society is declining due to the influence of *Gestell*. First natural resources, instead of having a meaning and a value of their own, are reduced to *Bestand* for humans, but eventually also us, the humans who perform this reduction through our development and employment of technology, start to treat each other the same way, rendering the human race itself *Bestand*. But why would we do that? Because with the specific revealing that comes with modern technology there follows a *concealing* that leads to a double forgetfulness. We forget to ask questions and come to accept that efficiency is best (it *is* more efficient, right?), and we forget that things, the world, could be revealed in different manners. Modern technology means not just a revealing of the world, but it is at the same time an all-encompassing concealment making us blind to other possible ways of revealing, other possible ways of being-in-the-world (Heidegger, 1977, 33f).

Heidegger's dystopian view of modern technology has been dismissed as irrelevant for postphenomenology (Ihde, 1979; 2010). The main reason for this is that it leaves no room to appreciate the specificities of technological mediations; it does not seem likely that wildly different technologies push us towards the same ontological and societal state. However, dismissing Heidegger's analysis only means dismissing a particular assumption about the human-technology relationship in the revealed world ("we are subordinate to technology"); it does not require us to dismiss the basic sense of technological mediation as a revealing-concealing structure, perhaps closer to how it appears in *Being and Time*. Accepting that the revealing-concealing structure of technologies constitutes the world as *a* world does not commit us to follow Heidegger into his brand of technological essentialism.

## THE EPISTEMOLOGICAL DIMENSION OF TECHNOLOGICAL MEDIATION: MAGNIFICATION-REDUCTION

In the epistemological dimension, technologies exhibit what we looked at introductorily, namely a *magnification-reduction structure*. Arguably, this is the type of technological mediation that has preoccupied Don Ihde the most (Verbeek, 2005). The concept is also his, although he first employed the less agreeable phrase sensory-extension-reduction (1979). Given Ihde's interest in and numerous investigations into technoscience, it is not so strange that Ihde has spent much time on the relation between technologically mediated perception and how this corresponds to how knowledge is structured.

As we saw, Ihde argues that any piece of technology that augments and strengthens our perceptual capabilities will simultaneously reduce or weaken other aspect of our experiential presence (1979, 9). Being intentionally directed at aspects of the world through a technology transforms the intentional experiences in accordance with characteristics of the technology. Had the directedness been done using a different technology—or none at all, we would have had a different experience. This holds not only for technoscience, but characterizes all technologically mediated experiences. Think about the experiential differences of feeling a blackboard using a dentist's probe instead of our fingers. Drawing a finger over the blackboard, we will feel that it has a certain temperature; we will feel certain dryness from the old chalk dust, and so no. With the probe, however, we cannot feel temperature or dust, but scratches and marks in the surface appear instead: "A microscopic presence is amplified through the probe thus extending my experience of the board to a level of discernment previously unnoted" (Ihde, 1979, 9). Amplifying or magnifying some experiential aspects through technology, inevitably leads to loss of experiential "access" to other aspects.

Reduction in technological mediation should not be thought of as impoverishing the perception. Furthermore, it should not be thought of as impoverishing technologically mediated knowledge; as rendering it relativist. Most people gesticulate when they speak on the phone. Not being able to "transfer" such gests, is that a blatant loss for communication? Of course not, the phone enables us to speak to persons not present; even though it is unable to convey the bodily gestures we often depend on in face-to-face conversation. The telephone does retain intonation, though. This, however, is lost in other forms of communication technology, such as email and twitter, which affords communicating in a different way from the phone or face-to-face, but have their own reductions. Combining several technologies, as in video-skyping, involves smaller "loss," allowing several sense modalities to be enhanced, but for instance smelling remains left out. It might be taken in in a future update of Skype or a similar platform for video conversations, though—it might prove helpful for, for instance, healthcare where nurses can gain a fuller picture of the condition of a patient they converse with. Is it at all possible to develop augmenting technologies that retains a "complete" experience, while enhancing some aspects of the world? However, why would we want that? Enhancing phenomena in the world is done because we want to focus on some particular aspect. Not letting this being accompanied by a corresponding reduction of other aspects will create noise and impede perception, of the scientific type as well as the everyday type.

As we saw in the previous section, technologies "open up new ways for reality to manifest itself" (Verbeek, 2005, 134). Although a technological mediation *presents* or *represents* an object by technical means, *manifestation* is a more fitting concept because presentations of any object will always correspond to a *perspective* on this object, relative to both the situated requirement—the task at hand—and to the material properties of the involved technologies. Through technologies with other qualities, or alternative contexts, or through users with other preferences and aims, the technological mediation would have "presented" the object differently but not for that reason necessarily more or less accurate; it would have been a *different* way for the object to manifest itself, suited to the particularities of that mediation.

However, this is not to say that the magnification-reduction structure of technology inherently relativize knowledge; the operation of a simple X-ray apparatus can exemplify this pretty well. An X-ray image results from the interaction of a body, an X-ray technician, the X-ray technology and the practice of using it. Decisive aspects here are, for instance, why an X-ray is taken at all, the function of the X-ray equipment, whether the technician can operate the equipment properly, and so on. The X-ray image produced, then, is a certain manner in which the body manifests itself. It is not the "truest" or "most objective" representation of the body, but is a *functional mode* of the body that reveals certain aspects deemed relevant for the task. The representation can hardly be called a pure objective depiction of the body, dependent as it is on the specific technical constraints of the apparatus, but it would likewise be absurd to regard it as an arbitrary construction, as it is clearly constrained by bodily properties. The technology focuses on and enhances, augments and translates certain aspects of the body, while at the same time it plays down or ignores others. In the terminology of the previous section, it reveals certain aspects of the body and at the same time, it conceals other possible representations of it. Technician, technology and body are in this case mutual constraints. To look at a body through X-ray's does not produce an arbitrary image, and it does not invent something that is not there, but it *projects a certain functional perspective* on the body. The representation is *relative to* X-ray technology, the competent use by an X-ray technician, *and the body itself*, but the knowledge that can emerge from it is not relativist. The X-ray image is, in this sense, an interpretation even prior to what we normally would label the interpretation of an X-ray image, and as such typical of technological practices of all kinds.

The magnification-reduction structure that is characteristic to technological mediations, then, is the way we *can* become acquainted with something; we are *able to* approach something only in as much as it stands out from that which it is not. This does not imply relativism, and as such it follows that magnification-reduction should not be seen negatively, as becoming the "only extent" that we can know something—"we can never know the thing itself"—but in the positive way as *the condition for us to know anything at all*.

It follows from this that the claim that a scientific image, fact, or model represents objective reality or conversely, that it only presents a constructed reality are arguments that are performed over a set of shared, but ill-conceived presuppositions. If something is said to be a mere constructed reality, this presupposes that there is something called reality somewhat behind or beyond it. Instead, regarding technologically mediated representations as "manifestations" means that the magnification-reduction structure of technologies is a way of *making reality*. Use of technology should not be seen as either granting us "direct access" to a pure nature-in-itself, nor should it be seen as putting a veil over this nature-in-itself. There is no such thing as a reality that we, by constructing the proper means, can describe in "pure" objective terms. The technological mediation is for this reason not a reduction or a diminished form of reality, but is itself reality, however, without the illusion of being an "objective truth" about reality.

## THE PRACTICAL DIMENSION OF TECHNOLOGICAL MEDIATION: ENABLING-CONSTRAINING

In our daily lives we depend heavily on all sorts of technologies to help us through both work and home life; from kitchen utensils to high-tech automatic communication systems. Rarely do we stop and think about these technologies as *constraining* our movements, rather we think of them as *enabling* them. And why should we not think that way? After all, for the most part we choose the technologies that we surround ourselves with precisely because we want them to help us achieve goals that we cannot make, at least not without much more effort, without said technologies. For patients living at home dependent on various forms of medical technology in order to ensure safety, to augment failing capacities, for treatment, and for social contact, the technologies certainly seem like *enabling* them to live at home rather than in a nursing home. A GPS device can make sure that the Alzheimer patient can move around outside without becoming lost; an iPad can be both an external memory device and a device for keeping in touch with friends and relatives; and telecare technologies can make it possible for patients to have a chronic disease cared for in familiar surroundings rather in a hospital or clinic.

However, also here technological mediation, in what I call the practical dimension, displays a two-sidedness that mirrors the two previous dimensions. In this dimension, technologies show an *enabling-constraining* structure that has a shaping impact on our behavior and actions. While enabling us to do specific things, technologies simultaneously shape *how* we do these things, and thereby divert out attention from other possible ways of doing it. In using a piece of technology we adapt to both a material and social reality that governs this device; we comply with the *affordances* of an item.

In being material, any technology affords certain uses, and for this reason, technological mediations are constrained by the affordances of a technology. This concept is primarily associated with James J. Gibson, who coined the term to denote the value or meaning a phenomenon in our surroundings has to us. Not thinking specifically about tools and technologies, Gibson claims that we primarily know objects, events, places, other animals and artefacts from what they afford us (1982, 404).[1] A handle, for instance, affords grasping, a sufficiently small stone affords throwing, and a sharp object affords piercing. Viscous substances afford being smeared while liquids afford pouring, and in certain cases drinking. Nutritional *and* poisonous substances afford eating, and so on (Gibson, 1982, 405). A mailbox affords letter-mailing; a social meaning, but materialized in the very design of the mailboxes (Gibson, 1979, 139).

Affordances are in a strong sense related to an item's materiality. However, without social belonging, technologies would conceptually be on par with found items. Because of this, we cannot regard affordances as merely expressing materiality or as belonging inherently to an item. On the other hand, technologies' affordances will in many cases surpass their conventional functions. Donald Norman points out that although "all screens within reaching distance afford touching, only some can detect the touch and respond to it. Thus, if the display does not have a touch-sensitive screen, the screen still affords touching" (1999, 39). However, given that the technical requirements are in place, un-thought-of affordances can become the norm for how technologies are used within a practice. Typewriters, for instance, were initially devices to assist those who had trouble writing for physical reasons, such as the blind (Tenner, 2003, 193). Affordances should therefore not be taken for merely being a social meaning. A physical mailbox might be discarded and thrown in the backyard, but it still affords putting mail into (although, if one wants the letter to arrive to the addressee, one is best advised not to do it).

What a piece of technology affords then is related to several factors. The materiality of the technology itself is important, but so are the abilities, competence, techniques, and perspective of those perceiving and interacting with it as well. Extending this into the social sphere; the role, function, and meaning the technology takes on in a society is to a large part dependent on the perspective and role it is perceived as having within this society. That a technology has a specific function within a social setting, however, does not prevent it from having a slightly or even vastly different function within another social setting—in a different culture, or attaining a revised function as time goes. "In both structure and history," Ihde says, "technologies simply can't be reduced to *designed functions*" (2002, 106). Instead, technologies are *multistable*; a piece of technology can take on different functions and meanings within different social settings (cf. Ihde, 2012). However, as I have argued more fully elsewhere, multistability is principally a descriptive concept (Kiran, 2012a), and in order to grasp the socio-technical dynamics behind multiple stabilizations of a material item, the concept of affordance proves helpful.

In enabling an item to be handled in specific ways, then, affordances shape our actions and behavior. Although affordances do not determine behavior they do constraint how our handling of items is performed. The concept of affordance reveals how technologies have both an *actuality* and a *potentiality*. This might seem like another pair of dual concepts, but these two concepts are for now best thought of as specifying in what manners technologies *enable*. The actuality of a piece of technology relates to how it is being used at a given moment, but it also denotes its social function, its conventional use; how a piece of technology usually is used within a practice. A technology's potentiality, on the other hand, covers various forms of unconventional use. For instance in the sense of single disconnected actions that a technology was not designed to perform, such as using the camera on our mobile phones as a temporary memory aid in order to remember where a bookshelf can be retrieved when we are at IKEA. A potentiality can also emerge when the meaning a technology has within a society changes and stabilizes in a new form, for instance when the conventional use of a small blue pill does not become to combat high blood pressure but to enhance erectile function. Or in cases of technology transfer when a technology, but not its socio-technological belonging travels from one culture to another, as in the case where Australian gold miners in New Guinea in the 1930s left behind cans of sardines, which were adapted into the culture of the natives as "centrepieces of the elaborate headwear they wore for special occasions" (Ihde, 1990, 125).

However, actualizing a technology's potentiality can take on another form that better accentuates the enabling-constraining duality of the practical dimension, namely *taken-for-grantedness*. We depend on many of the technologies that surround us, but of course we do not use them all the time. Still, there is a sense in which they still have a presence for us, even though they are not in use. Because we know that they have a potential use for us, we plan our actions and we organize our daily lives in accordance with technologies that we can engage for various kinds of support: we can buy more than we can carry because we have a car; we can plan on driving through a big city because we have a satellite navigation system; we become just a little slack in maintaining an appointment because we can send a text message to inform about and excuse our delay, and so on. Consequently, through their taken-for-grantedness, technologies shape our current activities even if their employment is a "mere" potentiality (Kiran, 2012b).

We find potential for several actions in a technology, but of course not just any kind of action. For instance, fine-grained brain surgery is not a potentiality of a hammer; to slap someone in the face is not a potentiality of Facebook no matter how many downward smiles one writes. Technologies in their actuality indicate their own potentiality. For instance, what problems are there with an item; what can be done better with it, and what should be done to its negative sides? Once a piece of technology has been articulated—maybe as a proto-type, or only as a plan, or even when it has been unleashed on society—we can start tinker with it. In this sense, materiality, and how it adapts to and is adopted in a practice, enables specific socio-technological developmental trajectories (Kiran, 2012b). In enabling specific trajectories, *how* a specific potentiality has been actualized constrains socio-technological developmental trajectories. It might even conceal alternative socio-technological developments (although rarely as overshadowing as Heidegger thought).

The dynamics between actuality and potentiality, then, displays the possible tasks we might perform; possible goals we might set; possible problems we might solve; and possible relations we might enter are very much related to the technologies we have in our surroundings, and *how* they are in our surroundings. In this sense, the practical dimension of technological mediations defines for us our *practical space*. For short, practical space is the horizon of our action potential.[2] In a line of thought well known from phenomenology, subjectivity emerges from embeddedness. Our awareness of ourselves is intimately connected to our *engagement* in our surroundings, but equally important is that our awareness of our surroundings is intimately connected to *our* engagement in them. This implies a co-dependency between subjectivity and objectivity; a subject is not a *mere* subject and the world is not a *mere* objective world—we are specific subjects and "our" world is a specific one.

Here, we begin to see how the practical dimension of technological mediations shapes subjectivity in a constitutional manner (this is more fully explored in Kiran, 2012b). Furthermore, the shaping of behavior that characterizes the enabling-constraining structure in some ways mirrors the revealing-concealing structure found in the ontological dimension of technological mediations. We saw above that affordances shape the very action that is actualized. When a particular action with a technological item is a conventional use, alternative uses of it is concealed, if only ever so little; for instance, we easily discover that a book can be used for holding another book open. Also in relation to taken-for-grantedness is it legitimate to talk of a concealing of possible actions. When we take specific technologies and their potentialities for granted in how we organize our lives, planning and executing daily chores and regular actions are shaped in manners that bring to light technological mediations that for us are well-known, tested, and tried. Habits conceal alternative behavior. Stating this more abstractly, we can say that actuality *reveals* potentiality.

## THE ETHICAL DIMENSION OF TECHNOLOGICAL MEDIATION: INVOLVING-ALIENATING

The ethical implications from the practical dimension are quite easy to spot. The kinds of behaviors that are being enabled and those that are constrained have ethical values attached to them, and has as such been widely discussed in the philosophy and ethics of technology. An affordance of mobile phones is that it makes it easier to cheat on one's wife. Most people, I suspect, will find that actualizing this potential of the mobile phone to be immoral. A technology that extracts oil and produces gas enables ease of traveling and transport and leads to economic prosperity in a society, but at the same time leads to higher emission of $CO_2$ and will for many be seen as being of questionable ethical value. On the other hand, technologies that constrain certain actions can have a positive ethical values attached to them, for instance when a speed bump hampers high speed driving past on a university campus (Latour, 1994, 38). And of course, technologies that constrain actions might be of negative value as well, while technologies that enable actions can have a positive value attached to them.

In the ethical dimension of technological mediations, a technology might often pull in different moral directions. Here, more than in the previous sections, can the duality of technological mediations be said to have certain value-laden ambivalence to it, which often means that we have to assess whether one side of the duality is positive and the other negative. The main issue, though, in a methodological sense, is to recognize that such moral ambivalence very often is related to the two-sidedness found in the other dimensions of technological mediation; as such, moral ambivalence will often be hard to bypass when developing and implementing new technologies. We might want our technologies to be morally unambiguous, but that might prove very hard to attain.

In the other three dimensions, there is always a constitutional element to technological mediation—whether it is the world as *a* world (ontological dimension), knowledge as perspectival (epistemological dimension), or actions as afforded (practical dimension). Accordingly for the ethical dimension; in opening us up to ethical dilemmas and issues, technologies constitute situations as *ethical* situations, with specific limitations on how the ethical issues are formulated (or formulate-*able*), and they constitute actors as ethical actors, with specific restraints on how we can behave and choose (Verbeek, 2011). This, I believe, makes the ethical dimension a dimension of its own, rather than merely being an aspect of the other there dimensions. And as such, very often we will find that in constituting an ethical situation, or an ethical actor, technologies display yet another two-sidedness; they can *involve* or they can *alienate*.

Ethics of technology is sometimes taken to be a critique of technology (Swierstra, 1997), aiming at assessing a piece of technology as either "good" or "bad," with the implication that bad technologies should be faded out or not be developed at all. The application for such dichotomies, however, is for many technologies arguable, as they harbor both good and bad sides. Furthermore, when it comes down to it consumers and regular users of technologies rarely have much of a say in whether technologies should be implemented in practice even if it is deemed to be a bad technology. The many control systems implemented as a result of new public management thinking is seldom liked by anyone having to use them, but they have to use and relate to them nonetheless. In assessing technologies ethically we need to go beyond the dichotomy, and rather assess them in terms of the opportunities and hindrances they pose for us to create for ourselves a good life, or, more tangible; a good work life, good marriage, good parenting, good care, and so on.[3] For many technologies, we cannot but relate to them on some level; an ethics of technology that assesses their *involving-alienating structure* can provide a helpful perspective on how to create a life *with* technologies (Kiran and Verbeek, 2010; Verbeek, 2011). Methodologically, this perspective also broadens the basis for traditional technology assessment (Kiran, Oudshoorn, and Verbeek, forthcoming).

The two-sidedness of the ethical dimension of technological mediation is operative in relation to information and communication technologies, and certainly discernable in those that are being used in healthcare. Technologies for monitoring a patient's illness are a good contribution to this person's life as a patient; telecare technologies, for instance, monitor the illness in a manner that enables the patient to stay at home rather than being in a hospital (Oudshoorn, 2011). These technologies are *involving* in that they grant patients an opportunity to construct an existence as a patient in familiar circumstances. However, the same technologies can be alienating, as they are a threat to the patient's privacy in affording surveillance, both in terms of the home being invaded by medical technology, turning the home into an outpost clinic (Oudshoorn, 2011, 173), and in terms of concerns about the safe storage and use of the accumulated data about a patient (Bharucha et al., 2009).

Relatedly, assistive (in Norway and Denmark sometimes called welfare) technologies are often regarded as granting autonomy and quality of life, as they ensure that the elderly and frail can stay at home rather than moving into nursing homes when their overall health start to deteriorate (Zwijsen et al., 2011). As such, also they are *involving* technologies, allowing users to create a better, and more dignified life with them than without them. However, assistive technologies also imply alienating aspects. For instance, some care receivers report that they feel controlled and dehumanized when living in a smart house (Astell, 2006).[4] Some users of fall detection technology would have preferred it if nobody found out that they had fallen, because they want "to cope on their own and contact the warden themselves only in extreme circumstances" (Brownsell et al., 2000). Assistive technologies such as a robot can be perceived as obtrusive, as disrupting a person's daily life (Faucounau et al., 2009). Many assistive technologies are meant to enhance independence, but in doing so, they also harbor the danger of leading to less human contact for the care receivers (Chapman, 2001).

Rather than enhancing autonomy, quality of life, and self-esteem, then, assistive technologies might have the opposite effect. Again, though, how the technologies are received and perceived by one person or one group of users depends very much on those persons and their situation; the technologies neither determine the involving aspect, nor the alienating. The important thing is to recognize that technological mediations can pull in both directions.

Additionally, we have technologies that on the face of it might be regarded as being *paternalist*; technologies that are meant to steer our behavior so that it emerges as moral behavior, but without involving us morally at all. Speed bumps are mentioned; we do not have to have a conscious wish to be moral, but we are as long as we slow down when going over one. Our wish might be to not damage our car, but the result is moral behavior (many ethicists would dispute this, claiming that without a conscious wish to be moral, no behavior can be labeled moral). Another example of a paternalist technology is the image of a fly that is etched onto the urinals of Schiphol airport. This has been done because it reduces spillage (it is said to reduce spillage with 80 percent); men tend to urinate with more precision when they have a target (Thaler and Sunstein, 2008, 4). This example, of a technology that is designed to *nudge* users into behaving in a specific way, is only one of several strategies to shape behavior through technology (Verbeek, 2011, 153; Tromp and Verbeek, 2011). Nudging is a gentle form of shaping our behavior; the speed bump is a bit more coercive, but we can choose to ruin our car by ignoring it (we would simultaneously break the law as there usually is a speed limit that goes with the speed bump). A car that does not start unless the driver puts on the seatbelt, a short-lived attempt from the 1970s to enforce seatbelt use, is an example of a far more forceful impetus. Here our behavior, if we want to take a ride in the car, is decided for us. For all these examples, there is a strong link between how a technology is designed, the effects it is intended to have on behavior, and the subsequent moral status of this behavior.

In a general sense, even though not all of these examples directly concern moral behavior (can accidental spillage of urine in a public toilet qualify as immoral?), paternalist designs breach autonomy (although that depends on how one define autonomy, cf. Verbeek, 2011) and might for that reason seem alienating, but they simultaneously are involving technologies as well, as they lessen the risk of everyday conflict and ensure societal co-existence. However, designing *for* specific behaviors, persuasively or coercively, is not uncomplicated. As we saw above, technologies are multistable, and as I have argued extensively elsewhere; thinking that a given behavior, moral or otherwise, can be ensured through designing and developing technologies in specific ways is to underestimate the material, social, cultural, and personal complexity involved in how technologies are received and in how technological mediations are performed (Kiran, 2012a).

## CONCLUDING REMARKS

I am not claiming that this is a comprehensive overview of the many dimensions of technological mediation. Far from it, I can for instance think of a *political dimension* as well, perhaps with a *liberating-oppressive duality* to it. And maybe someone would like to emphasize technological mediations' *communicative dimension*, which might display an *attaching-detaching duality*. Furthermore, there could be a *legal dimension* where technological mediations *allow and prohibit*. However, I have discussed the above four dimensions because I found them sufficiently different, yet interweaved in many respects. Other dimensions, I suspect, can often be analyzed as either falling in under one of the above (the communicative as subsumed to the practical dimension) or as being a combination of two or more of the above (the political as belonging to both the practical and the ethical dimension). However, I am not implying that nothing fruitful can come out of investigating them in their own right.

My principal aim with this paper has been to disclose how the four dimensions, taken together in all their diversity, delineate technological mediation as a significant constituent element of our lives, identities, and of our lifeworld. Seeing as this concept is already indispensable for postphenomenology, exposing and explicating the two-sidedness of these dimensions substantiates the need for postphenomenological studies of technology-in-practice, in (techno-) science as well as in other professional and non-professional occupations; from healthcare to education to dwelling.

However, that is not to say that the concept of technological mediation has little or no value outside postphenomenology, and a further aim is that this analysis goes some way in indicating how the concept can be methodologically beneficial for research in a number of scientific disciplines; not just technology studies and technology assessment, but in sociological, anthropological, and historical analyses of our technologized society as well as varieties of philosophy, psychology, and ethics that in some way deal with technological cultures and innovation in empirical practices. That being said, the concept should not necessarily take center stage, but as the various kinds of two-sidedness expose the influence of technology on how actors, situations, and cultures are being shaped into how and what they are and on socio-technological development, technological mediation should be valuable in the effort to define and demarcate fundamental concepts in the methodology of several disciplines.

## ACKNOWLEDGMENTS

The research behind this article was funded by *The Norwegian Research Council* (Program: Praksisrettet FoU for helse—og velferdstjenestene (PraksisVel); project "The Medical Home—Sustainable Services and Technology for Home Medication").

## NOTES

1. Gibson mainly discusses what "affordance" means for a *visual* psychology. In line with a more general sense of what it means to afford (Norman, 1999; Hutchby, 2001), I refer to how an item affords *actions*, not merely interpretations.

2. The concept is inspired by Maurice Merleau-Ponty's notion (1962, 137ff), who employs it to convey that how we are in the world does not come down to a mental representation of an objective world (called mental space), but is more importantly related to our body and its motility. My usage is related, but a bit different.

3. This is not to say that technologies or technology developing processes (or science) cannot be assessed as "good" or "bad," merely that that is not the kind of ethics of technology that I am preoccupied with here.

4. Smart houses can have various devices: bathtub monitor, automatic switcher for the oven, locating devices (to find, for instance, mislaid keys), lights that turn themselves on and off depending on whether the person living there is in or out of bed, and message boards around the house that display all sorts of messages (Astell, 2006).

## REFERENCES

Astell, A. J. (2006). "Technology and Personhood in Dementia Care." *Quality in Ageing* 7 (1): 15–25.

Bharucha, A. J., V. Anand, J. Forlizzi, M. A. Dew, C. F. Reynolds, S. Stevens, and H. Wactlar. (2009). "Intelligent Assistive Technology Applications to Dementia Care: Current Capabilities, Limitations, and Future Challenges." *The American Journal of Geriatric Psychiatry: Official Journal of the American Association for Geriatric Psychiatry* 17 (2): 88–104.

Brownsell, S. J., D. A. Bradley, R. Bragg, P. Catlin, and J. Carlier. (2000). "Do Community Alarm Users Want Telecare?" *Journal of Telemedicine and Telecare* 6 (4): 199–204.

Chapman, A. (2001). "There's No Place Like a Smart Home." *Journal of Dementia Care* 9 (1): 28–31.

Faucounau, V., Y. H. Wu, M. Boulay, M. Maestrutti, and A. S. Rigaud. (2009). "Caregivers' Requirements for In-Home Robotic Agent for Supporting Community-Living Elderly Subjects with Cognitive Impairment." *Technology and Health Care: Official Journal of the European Society for Engineering and Medicine* 17 (1): 33–40.

Gibson, J. J. (1979). *The Ecological Approach to Visual Perception.* Hillsdale, NJ: Lawrence Erlbaum.

Gibson, J. J. (1982). *Reasons for Realism: Selected Essays of James J. Gibson*. Hillsdale, NJ: Lawrence Erlbaum.

Heidegger, M. (1927/1962). *Being and Time*. Oxford: Basil Blackwell.

Heidegger, M. (1969). *Discourse on Thinking*. New York: Harper & Row.

Heidegger, M. (1977). "The Question Concerning Technology." In Heidegger, *The Question Concerning Technology, and Other Essays*, 3–35. New York: Harper & Row.

Heidegger, M. (1988). *The Basic Problems of Phenomenology*. Bloomington: Indiana University Press.

Hutchby, I. (2001). "Technologies, Texts and Affordances." *Sociology* 35 (2): 441–456.

Ihde, D. (1979). *Technics and Praxis*. Dordrecht: Reidel.

Ihde, D. (1990). *Technology and the Lifeworld. From Garden to Earth*. Bloomington: Indiana University Press.

Ihde, D. (2002). *Bodies in Technology*. Minneapolis: University of Minnesota Press.
Ihde, D. (2010). *Heidegger's Technologies: Postphenomenological Perspectives*. New York: Fordham University Press.
Ihde, D. (2012). *Experimental Phenomenology: Multistabilities*, 2nd edition. Albany, NY: SUNY Press.
Kiran, A. H. (2012a). "Responsible Design. A Conceptual Look on Interdependent Design-Use Dynamics." *Philosophy & Technology* 25 (2): 179–198.
Kiran, A. H. (2012b). "Technological Presence: Actuality and Potentiality in Subject Constitution." *Human Studies* 35 (1): 77–93.
Kiran, A. H., and P-P. Verbeek. (2010). "Trusting Our Selves to Technology." *Knowledge, Technology & Policy* 23 (3): 409–427.
Kiran, A. H., N. E. J. Oudshoorn, and P-P. Verbeek. (forthcoming). "Beyond Checklists: Towards an Ethical-Constructive Technology Assessment." *Journal of Responsible Innovation*.
Latour, B. (1994). "On Technical Mediation—Philosophy, Sociology, Genealogy." *Common Knowledge* 3 (2): 29–64.
Merleau-Ponty, M. (1962/1945). *Phenomenology of Perception*. London: Routledge.
Norman, D. A. (1999). "Affordances, Conventions, and Design." *Interactions* 6 (3): 38–42.
Oudshoorn, N. E. J. (2011). *Telecare Technologies and the Transformation of Healthcare*. Basingstoke: Palgrave Macmillan.
Swierstra, T. (1997). "From Critique to Responsibility." *Techné: Research in Philosophy and Technology* 3 (1): 68–74.
Tenner, E. (2003). *Our Own Devices. The Past and Future of Body Technology*. New York: Alfred A. Knopf.
Thaler, R. H., and C. R. Sunstein. (2008). *Nudge: Improving Decisions about Health, Wealth and Happiness*. New Haven: Yale University Press.
Tromp, N., P. Hekkert, and P-P. Verbeek. (2011). "Design for Socially Responsible Behaviour: A Classification of Influence Based on Intended User Experience." *Design Issues* 27 (3): 3–19.
Verbeek, P-P. (2005). *What Things Do. Philosophical Reflections on Technology, Agency, and Design*. University Park: Pennsylvania State University Press.
Verbeek, P-P. (2011). *Moralizing Technology. Understanding and Designing the Morality of Things*. Chicago: The University of Chicago Press.
Zwijsen, S. A., A. R. Niemeijera, and C. M. P. M. Hertogh. (2011). "Ethics of Using Assistive Technology in the Care for Community-Dwelling Elderly People: An Overview of the Literature." *Aging & Mental Health* 15 (4): 419–427.

# 3

# Postphenomenological Cases

## *Chapter Eight*

# Tracing the Tracker

## *A Postphenomenological Inquiry into Self-Tracking Technologies*

Yoni Van Den Eede

This chapter sets out to investigate the emerging phenomena of self-tracking and digital self-tracking technologies through the lens of postphenomenology. By putting postphenomenological notions to work, it will appear, we can start to "trace the tracker," although that might require some fine-tuning of the postphenomenological framework itself, in turn.

Self-tracking technologies are all around us; in multiple ways. They are growing in popularity as more and more devices find their way to the market and consumers. But they are also literally around us, as a prime instance of wearable tech, residing by definition close to a user's body. Products such as those developed by Jawbone, Fitbit, and BodyMedia are designed to track one's everyday activities and collect and store data about them in the form of for instance steps taken, calories burned, distance traveled, and sleeping patterns. The tools generally consist of on the one hand a wrist, arm, or headband, or any other measuring artifact containing sensors (e.g., accelerometers) that a user has to wear or carry on one's body, and on the other hand a mobile or online app through which one can consult the "result" of one's measuring activities.

Self-tracking, also known under the rubric of personal analytics, personal informatics, and quantified self, is done mostly in function of optimizing one's "performance," be it in sports, professionally, or more broadly speaking in terms of well-being. It is also deployed in a more specialized vein in the context of medical treatments that require a quasi constant monitoring of certain physiological variables, as for example in the case of diabetic patients. Of course, the activity is not wholly new to the extent that people in contexts like these have always sought to improve their condition by keeping track of variables of all sorts (Schüll, 2012). But as a plethora of devices is now becoming available that enable the accumulation of heretofore inaccessible data (e.g., brain activity), easier storage, aggregation of different data streams, analysis and display of data by way of algorithms, et cetera, we may speak of, in postphenomenological terms, an intensified technological mediation of the "generic" act of monitoring and tracking oneself.

As the phenomenon is gaining a strong foothold in everyday life, the scholarly research on it, too, has been steadily growing. I have presented an overview of available literature elsewhere (Van Den Eede, 2014); here I seek to merely probe the activity of technologically mediated self-tracking, from a fresh perspective,[1] namely, by engaging a couple of central postphenomenological concepts, most of them hailing from Don Ihde's work. Two characteristics of self-tracking make the postphenomenological conceptual toolbox particularly well-suited for this task. First, self-tracking is all about the enmeshment of bodies, technology, and perceptual experience. This links up neatly with postphenomenology's attention to human-technology-world relations and to embodiment. Second, as self-tracking technologies in their current form are still relatively new, it may be sometimes hard to see beyond the image projected of them by marketing and by explicit or implicit cultural presuppositions. The postphenomenological "method," bent as it is on holding up a phenomenon to the light of possibility, and serving in that way as a hands-on tool for the sober but open-minded investigation of that phenomenon and its more or less outspoken "promises" in the "here and now," can compensate for this potential shortsightedness.

The chapter unfolds into two large parts. In the first, I endeavor to frame self-tracking through the lens of some of the most important postphenomenological notions. The second part digs deeper into the results of that first inquiry in order to point out (1) how research on self-tracking technologies can benefit from postphenomenological insights and (2) how, conversely, some of postphenomenology's concepts should be complemented on the basis of what can be learned from self-tracking.

## SELF-TRACKING SEEN THROUGH POSTPHENOMENOLOGICAL LENSES

In Ihde's work, multiple concepts have been developed that may help us scrutinize digital self-tracking technologies, first and foremost by viewing them in another, unusual light. I will list these notions and consider how they apply.

## Perception, Embodiment, Field-Fringe Relation

The postphenomenological corpus has its roots in classic phenomenology, which starts by attending to the experiences given to a subject. From there, the subject reflexively finds itself (cf. Ihde, 2012, 24ff.). This idea seems well-suited to self-tracking. Tracking oneself may in a sense be regarded as some kind of literal realization of this keystone phenomenological idea. In tracking one's physical exercise patterns, for instance, a data set about distance, location, speed, calorie burning, et cetera is constituted. In this constellation of data, then, the exercising subject may find oneself back. It is as if the tracking is there first, and only in a second instance the "self" comes about. Of course this is just by way of metaphor; experience in its fundamental phenomenological capacity is *always* primordial. Even before one observes any data, the "world" and the "I" have already been found throughout, for example, the running activity. But still, the added "data double" (Ruckenstein, 2014; Haggerty and Ericson, 2000) that the tracking brings about, may be seen to follow a similar pattern. One discovers oneself *through* the self-tracking technology—because one always discovers oneself reflexively.

This is interesting: it may serve, right away, as an admonition with regard to any possible utopian dreams that we may have about self-tracking, for example, that they may help us to disclose our true selves, or something of that order. For phenomenologically speaking, the subject is enigmatic from the start (Ihde, 2012, 11). Moreover, Ihde is there to remind us that perception is never passive; it is active and constructive (a.o., 2009, 62). Whatever window self-tracking devices may offer one on the world, the view is essentially constructed. Be that how it may, within perception we are able to discern some fixed characteristics. First, perception is always embodied (cf., Ihde, 1998; 2010). Even when the visual sense dominates, experience is still given to an embodied subject. This point, too, is well illustrated by self-tracking technologies, that almost literally confront us with our corporeal actuality. Here again, the tracked self may be seen to act as a sort of phenomenological "double" of our world-oriented but embodied being. Second, we can find "noematic structural invariants," aspects that belong to the way the world is organized so to speak. One of them is the relation between figure and ground or field, another one is that between field and fringe or horizon (Ihde, 2012, 40). No thing is ever perceived in isolation. One may focus on it, but it is always there in relation to a ground or field. We can, however, try to get that broader context in view. The field, in turn, frays out into a horizon, a fringe. Also of this horizon we may become aware, although often dimly and perhaps with great effort. The essential idea here is that we can study the complex of our experiences in function of this three-component structural relation. We may for instance categorize "material" elements of self-tracking technologies—sensor, app, …—along this spectrum, and see moreover how their states vary across time and space.

## Human-Technology-World Relations

A more detailed exercise along these lines can be elaborated by utilizing Ihde's "phenomenology of technics," his outline of different kinds of human-technology-world relations. As is well-known, Ihde defines four of them: embodiment, hermeneutic, alterity, and background relations (1990, 72–118). Nothing prevents certain technologies to actually take part in all four; an automobile for example can be said to harbor elements from all relations. Self-tracking technology, too, seems to represent a fitting example of such multi-facetedness.

Let us take as an illustration a system consisting of a wristband with sensor, smartphone app, and online dashboard (such as the Jawbone Up[2] or the FitBit Flex[3]). Obviously the system is partly embodied: the wristband, at least after some time, gets incorporated into the body scheme just as much as glasses are. But the band may also have a hermeneutic aspect, if it spots a display, for example. Almost wholly hermeneutic, by contrast, are the graphical interfaces offered by the app and online tool: one reads the "world" through them. In this particular case we can ask, also in line of the above considerations, to what extent the world "read" is in fact the perceiving subject ("me") itself; I will elaborate upon this issue later on. The whole system as such, then, or parts of it, can also acquire the character of an alterity. One may start to regard the device, including its interfaces, as a kind of personal trainer. Companies play on this potentiality by building in goal-attaining or coaching functionalities. For instance, one may receive a weekly overview of one's progress, badges when certain goals are reached (e.g., number of steps, calories burned, …), or alerts when one is "too idle." The system thus manifests itself as a "third," spurring one on to take actions. Nevertheless, at times the whole tracking and/or coaching system may also disappear wholly from view and temporarily turn into a background phenomenon. When one does not notice the armband for example—due to its embodied character—and forgets about goals, numbers, et cetera for a while, self-tracking takes on an "invisible," background character. And in any case while in use and even if one experiences the device's "nudges" (Thaler and Sunstein, 2009), some parts of the system such as the online dashboard may for the largest part remain in the background. The question does remain, of course, in how far the device in that case does not work its influence in a sort of unconscious, unnoticed manner. Strictly speaking, however, this is of no concern to (post)phenomenology, which limits its research to the immediately given.

## Micro and Macroperception

That, however, does not mean that the immediately given is untainted by larger cultural contexts. In fact, Ihde has done a lot to show that the two are intimately interwoven. He terms the two levels micro and macroperception, respectively (1990, 29). Especially when it concerns self-tracking, the influence of the cultural context, that is, macroperception, is crucial. We cannot possibly investigate the phenomenon in isolation from our Western, (post)modern culture of fitness, physical competence, and performance; and we can well imagine that to other cultures, self-tracking devices may appear as alien as firearms were once to indigenous South-Americans. Most of these new devices are marketed as tools with which to improve one's performance or health. They would be hard to situate, let alone understand if not for this surrounding constellation of norms, habits, and expectations.

In this regard, an even more specific distinction that Ihde makes—actually as a variation upon the micro-macroperception dichotomy—is of importance: that between "body one" and "body two" (2002, xiff.). Body one refers to the body that we are in an embodied, experiential, emotive sense. Body two is our social and cultural body, the body that we are invited or pushed to shape according to cultural fashion, expectation, rules, and norms. And traversing both, Ihde adds, is a third dimension, that is, technology (or materiality). Especially in the case of self-tracking technologies, this image carries much pertinence. Anthropologist Minna Ruckenstein remarks: "Journal entries [of her tracker respondents] reveal the biosociality of occurrences and activities of everyday life. Bodily reactions cannot be separated from social relations, even if people wish they could be" (2012, 15). A lot then depends, at least for theory, on which account of body two one chooses to put forth: the active "sports body," as in Merleau-Ponty's case, or the "culturally fixed and acted upon body of Foucault" (Ihde, 2002, 26). This leads to the next point, namely, about multistability.

## Multistability

Multistability may be said to be the most important postphenomenological concept, not in the least because it ties together most other strands in the framework. In relation to the matter at hand, it must be asked: what sorts of multistabilities do self-tracking technologies harbor? Obviously, if one would query their developers as to what they are "for," one would probably receive answers congruent with what has been said in the preceding paragraphs: they promote and improve health, sleep, performance, and what not.

Keeping in mind Ihde's expansion upon variational theory, however, one would do best in tuning one's antennae to other possibilities as well. Self-tracking technologies may also serve as biopolitical tools, instruments to help keep subjects sufficiently in check and disciplined, that is, healthy, active, profitable (cf. Lupton, 2012). But alternative accounts, not so strictly early-Foucauldian in spirit, are possible too, according to which subjects actually creatively participate in their own "surveillance" (Albrechtslund and Lauritsen, 2013; Ruckenstein, 2014). Such accounts suggest that beyond issues of surveillance and biopolitics, more variations are conceivable. Other usages for self-tracking systems may be found.

This is to a certain extent illustrated throughout the Quantified Self (QS) community, that gathers people actively involved in self-tracking worldwide.[4] A lot of trackers are at heart driven by the health and performance-related motives to which the devices are in the first instance tailored. But some of them also gear the systems to other needs that do not immediately appear to fit the biopolitical agenda, or do so at least in highly ambivalent ways. One tracker (dos Santos, 2013) for instance monitors social relationships, in terms of among others time spent together and subjects covered in conversation. Admittedly, one could allege that this entails, what with the sheer reduction to numbers and statistics, a biopolitical colonization of one of the last refuges from the disciplinary network, that is, non-mediated, "natural" human interaction. But the person in question also reports spending more time with members of his family now because of the shock that he received when he noticed during tracking that he always felt rushed—as so many nowadays in social interaction do—in talking to his relatives, as in the course of every interaction always already something else was beckoning his attention. A similar phenomenon is reported by Ruckenstein who observes in her empirical research on the tracking of stress and recovery levels in relation to kinds of activities, how respondents actually found out that "non-efficient" activities such as idling, playing with the kids, et cetera actually gave them the most recovery, and this made them perceive those so-called useless pastimes in a new light: they appear to be "useful" after all. (Ironically this, of course, I would add, straight away subjugates them to discipline again.) Using a somewhat more macabre illustration, then, Ruckenstein remarks:

> Recontextualized data doubles might also argue against, ignore, or bypass normative notions of auto-correcting and transforming selves. For instance, anorectics participating in pro-ana online communities share weight-loss tips and offer "thinspiration" to involve and encourage others in body projects [...] which can take advantage of various forms of self-tracking, including self-monitoring devices, in order to be achieved. (2014, 71)

*Non-neutrality, amplification-reduction.* Be it how it may, self-tracking technologies, thus, could turn out to be quite something else than what their makers, marketeers, or users now mean them to be. Their possible variations or multistabilities will have to be manifested and explored through future usage. Different "trajectories," in Ihde's terminology, may become in time apparent—and these may refer to either the materiality of the technology, the bodily technique in using it, or the cultural context environing it (cf. Ihde, 2009, 18-19). Nonetheless, it remains the case that at any time, that is, in any one of such trajectories, a technology can exhibit certain "instrumental inclinations" (1993, 54), as Ihde puts it. It is non-neutral. More precisely, it can be seen to amplify or enhance something, and to reduce something else. Glasses for example, enhance eyesight but reduce, for one, motility and flexibility when doing sports. Put differently, technologies always bring transformations (Ihde, 1990). What transformations, then, do self-tracking technologies bring? Quite explicitly, these products are brought to market with the deliberate aim of effectuating, exactly, transformations: they serve to change one's behavior in such manner that one starts to live more healthily and efficiently—or at least so the marketing tale goes.

But what else do or could they "do"? They may bring about change for the better in one's exercise routines, but—on the reductive side—simplify rich, "felt" experience to stats and graphs. Just as, Ihde says, photography teaches a certain way of seeing (1993, 48), self-tracking technologies may teach a certain way of living, probably bent on performance, goal-attaining—not without a reason, some articles on the Quantified Self movement of late have the phrase "living by numbers" in their headlines[5]—and visualizable info. The activity of self-tracking is never neutral, but also never wholly deterministic either. Nevertheless, self-tracking technologies may, like so many technologies, fascinate us; and, Ihde remarks, fascination about technologies is usually a function or marker of their amplification (1990, 78). One should not forget that technology fascination often roots in some fantasy, "technomyth" (Ihde, 1998, 105), or "technology plot" (Ihde, 2012, 131), that for the most part resides, of course, in the macroperceptual cultural context. Here, in the case of self-tracking, that connection is extremely tight: what the devices do is, precisely, attempting to draw the microperceptual bodily activity closer to the macroperceptual cultural expectational horizon. As such, in a recursive maneuver, enchantment with them—Ihde: "fascination and adornment are as good a way into the hearts and minds of humans as any" (1993, 36)—and by proxy, with the sports and performance culture that breeds them in the first place, is a full-blown part of exactly their "instrumental inclination."

## Imaging Technologies, Instrumental Realism, Epistemology Engines

Finally, one last aspect of Ihde's thought needs to be discussed, one that may seem at first sight relatively unrelated to what has already been elaborated, but is in fact closely connected to it: his long-standing interest in imaging technologies and their attending epistemological issues. For what are self-tracking technologies else than a kind of imaging technologies? Especially Ihde's thinking in terms of isomorphic degree may be enlightening here (2009, 45ff.). In the history of scientific visualization, he discerns two revolutions. The first one, epitomized by the telescope, circled around greater and clearer magnification, but the image created by the technologies remained isomorphic with the original object. A telescope, for example, lets one see a planet "as if" one is close to it—the structural characteristics of the mediated and the unmediated image are the same. However, the second imaging revolution introduces non-isomorphism, as it brings in technologies such as radio telescopes, that receive signals from that part of the wave spectrum that is unperceivable to humans, converting these to perceivable images. Here, clearly, there is a structural difference between the original data and the constructed image. And due to the emergence of technologies such as computerized enhancement and contrastive techniques, the fact of construction is even more emphasized. With the second revolution, non-isomorphic in nature, we see the birth of a "second sight," "a translation into the visible of phenomena that lie beyond literal vision" (Ihde, 2002, 47).

All of this relates to Ihde's work in the epistemology of science. By way of the notions of "technoscience" (2009, 25ff.; Ihde and Selinger, 2003) and "instrumental realism" (1991; 1998), he elucidates how technological instrumentation has always had primacy over "pure" scientific abstraction, not the other way around (a state of affairs, he says, that Heidegger observes but that Husserl misses). "*[I]nstrumentation* transformed knowledge gathering and production in the sciences" (Ihde, 1998, 43, original emphasis). Therefore we need an "expanded hermeneutics" that accounts for the materiality of science as well. Since this materiality often takes the shape of visualizing, expanding hermeneutics likewise means: widening its boundaries until it includes not only textual interpretation, but also visual hermeneutics. Crucially—and this relates to a point already made about the embodiment of perception—the technologically mediated visual perception on which scientific practices are based, is grounded in embodiment just as much. "The embodiment of observers is [...] an invariant in science" (Ihde, 2009, 61). Scientific imaging practices are, though mediated, "fully multisensory and embodied" (Ihde, 2002, 59).

Although these considerations have grave ontological consequences—for instance with regard to the purported "symmetry" between humans and other entities (cf., Ihde, 2002, 67ff.; Pickering, 2006)—that cannot be worked out here, the relevance with regard to self-tracking may be clear. But there are complications. How should one frame, say, the isomorphism vs. non-isomorphism distinction? The "image" that self-tracking tools construct—for it *is* a construction, not a neutral or objective representation of an original situation—cannot be said to be wholly isomorphic with the "object" it reproduces, but it cannot be called completely non-isomorphic either. Take for example the counting and rendering of steps taken. One may walk, say, 8235 steps in a day. It may "feel" as if one has taken, indeed, about that amount of steps. Certainly when one is used to using a pedometer, one is able to grossly estimate the amount of steps based on a loose assessment of for instance time spent walking and locations visited. In that sense it could be said that some form of isomorphism is at play. Nevertheless, the rendition of the precise number—of course, all devices have a margin of error in this regard—is a new, constructed addition, since in "normal" conduct, no one counts his or her steps while taking them. Here, thus, some non-isomorphic constructive "translation" is going on. But the perception and interpretation of its results still feedback to one's embodied being, in this case even very actively, for the imaging *concerns* our embodied being. We perceive in an embodied manner an however objectified version of our embodiment.

The question remains what sort of knowledge exactly these mechanisms bring. As development in scientific knowledge follows from instrumentation, we can ask: what forms of knowledge—scientific or other—might self-tracking technologies, in time, produce? And, moreover, how will they influence what we actually conceive of as *pertinent* knowledge? Ihde's notion of "epistemology engines," at last, is of use here. Epistemology engines are devices "that bring human knowers into intimate relations with technologies or machinic agencies through which some defined model of what is taken as knowledge is produced" (2002, 69). Imaging technologies have been in this sense epistemology engines. Thus the camera obscura, for instance, Ihde argues, brought forth a certain obsession with vision. How will self-tracking systems affect our view of what worthwhile knowledge precisely implies? One thing is for certain: if a central postphenomenological claim is that "*only through being technologically mediated is the newly produced knowledge possible*" (Ihde 2009, 55, original emphasis), surely self-tracking can be seen as an actualization of exactly this claim.

## CONSEQUENCES OF/FOR POSTPHENOMENOLOGY: TRACING THE TRACKER

Up until now, I have scrutinized digital self-tracking technologies through postphenomenological lenses, that is, by way of central notions from the postphenomenological corpus. Let me now attempt to dig still somewhat deeper and outline a few important critical consequences of the foregoing investigations—first, of postphenomenology for the understanding of self-tracking, and second, vice versa, of self-tracking technology for the postphenomenological body of ideas itself.

I already remarked upon the multistability concept's importance for postphenomenology. Put even more strongly, however, postphenomenology as such might even be conceived of as nothing less than a *plea* for multistability or, more precisely, an openness to and search for multistabilities. Ihde demonstrates for instance how the history of imaging technologies (cf. supra), branching out into multiple "trajectories" (cf. also 2012, 131ff.), makes for an extensive illustration of multistability. One could say the same of the historical development of many other technologies, scientific models, theories, and what not. We can notice this, if we are attentive enough, even in our everyday lives. From all sides, we are bombarded with snippets of health advice, and frequently these appear to contradict some former research outcome. Take for example the evolution of nutritional advice throughout the decades. At one time in the recent past, saturated fats were seen to be great evil-doers, whereas sugars were looked upon with much less suspicion. By now, of course, sugars have fallen in grave disrepute, with an avalanche of diet or light products as a consequence. Obviously this is a highly simplified account of what is in reality at any time a complex interaction of technoscientific practices, market interests, cultural habits, and so forth (cf., Scrinis, 2013). But it serves to illustrate the possibility that in fact *all* phenomena carry with them different potential "stabilities."

This makes the study of self-tracking technologies sufficiently complicated. As said, micro and macroperception are intertwined. Not only should one expect the tracking technologies themselves to be multistable, also *what* they track, we need to remind ourselves time and again, is everything but carved in stone. And what they track, as was also seen, is co-framed by, in Ihde's words, "[t]he contemporary passion for exercise, body development, or retention of a youthful body" (2002, 25), and by the myriad palette of health recommendations issued on a steady basis by academic research institutes, pharmaceutical companies, governmental panels, and so on. Certainly in the case of self-tracking, in which bodily comportment, technological materiality, and cultural practices can be seen to be even more tightly knotted together—actually, that interwovenness, more than anything else, defines the whole activity—the potential multistability of *all* matters could—and should—constitute a constant guideline for the investigation. The designer fallacy (Ihde, 2002, 106; 2008, 19ff.), namely, the idea that technologies act strictly in manners that their designers have planned, might find a pendant in what we could dub a "fixed entity fallacy": the idea that phenomena can be reified as fixed entities with unambivalent qualities ("sugar" is "bad"; "exercise" is "healthy" . . .).

That, it could be pointed out, also has consequences for the "hacking" concept hailed by some in the Quantified Self community.[6] Hacking, in its most technical definition, means the search for and realization of applications for a device, namely a computer, that the developer did not plan or foresee. Nowadays the term has acquired a broader significance. Within QS to some extent the idea lives that one can hack one's own health, control one's well-being beyond the confinement of classic health models (cf. also Melanie Swan's expectation that QS might revolutionize health care as we know it [2009; 2012]). In principle one does so with the help of technologies, which in turn can be hacked and deployed for unforeseen purposes. In this sense, hacking becomes something of a lifestyle (and indeed terms such as life hacking, body hacking, mood hacking, et cetera are being used nowadays). Nevertheless, the above reflection on multistability should remind us of the possibility that the "hacking work" is never done. In fact, in hacking, one should be very wary of the phenomena that fall outside of the supposed circle of "hackable" things. Those, one surmises, may in the first instance concern ideas and ideals that tend to escape our conscious reasoning anyhow, our cultural expectations and norms regarding efficiency, performance, and physical condition.

From another perspective, then, one may also ask: could there be some advice here for design? Ihde argues, however specifically with regard to embodiment relations, but the point applies well to self-tracking technologies in whole: "embodiment relations can be said to work best when there is both a transparency and an isomorphism between perceptual and bodily action within the relation" (1990, 90). Two crucial matters stand out in this quote. First, transparency. Indeed, other authors point to the importance of "seamlessness" as well. Melanie Swan remarks that one of the barriers to widespread adoption of self-tracking technologies at the moment is practical: they are not "automated, easy, inexpensive, and comfortable" enough (2013, 93). But then there is Ihde to warn us about our perhaps unconscious wish for total transparency (1990, 75): wanting the benefits of technology without the disadvantages; the amplification without the reduction. This is an impossible wish: technologies always transform, no matter how small or trivial the affected areas.

Second, isomorphism, that was already discussed to some extent, and of which there is certainly no clear-cut account in the context of self-tracking. When one considers graphics of sleep patterns, step count, distance traveled, et cetera, of what exactly are those images? The term "self-tracking" alone might confuse us into thinking that the technologies offer us painless views, neutral windows on our "selves," since there is *some* degree of isomorphism. One for instance recognizes in the graphical representation of one's sleeping pattern the sleep experience, the "feel" of the previous night. We could be tempted into equating "world" with "human" in Ihde's schematic representation of, for one, embodiment relations (table 8.1). But, as said, even to ourselves "we" are not immediately given; the "world" has, phenomenologically speaking, primacy. What is more, self-tracking technologies take part in more than embodiment relations alone; they also constitute hermeneutic relations, of non-isomorphic character moreover, to the extent that the "data double" created is a construction, truly a *double*, and not merely a mirror of "world."

Connecting this to observations already made, and involving the other forms of human-technology relations too, we could state that ambiguities (or, distortions?) abound at at least three levels:

1. First person level: Technologies may be almost perfectly embodied, but the "I" is as such already only reflexively found.
2. Second person ("data double") level: Data doubles are constructions, translations, that need hermeneutic reading and deciphering.
3. Third person level: The "alterity" "trainer" or "coach" is a function of cultural norms, lifestyle choices, "body two" preconditions, et cetera…

And, one might even envision a fourth level, if one cares to follow Melanie Swan in her prognoses about the future self, inspired by Kevin Kelly, expecting the emergence of the "exoself." Such a self would be, due to the future abundance of self-tracking-like technologies, "spatially expanded, with a broad suite of exosenses" (Swan, 2013, 95). It could be, Swan adds, "a sort of fourth-person perspective" (Ibid., 96). Looking at these issues in a postphenomenological light, one could expect such fourth-person "presence" to mostly escape the awareness of an (individual) embodied self, and thus have for a large part the status of a background technology. Of course, then, it should be remarked, background technologies bring along transformations, ambivalences, and multistabilities just as much. Once again, not only are these issues to be dealt with by users; designers, too, might take up the challenge to creatively play on these complications.

**Table 8.1.**

(human-technology) → world
if world = human, then
(human-technology) → human (-world)
or, perhaps:
(human ↲ technology) (-world)

An analogy may be spotted with virtual technologies, that Ihde also treats of. Does self-tracking represent some form of technological virtuality? Ihde speaks of a "here-body" versus a "disembodied over-there-body" or an "image-body" (2002, 6). Of course the context here differs a bit from that of the issue at hand. But still, there is some similarity to the extent that, as Ihde argues, "in a broader, more phenomenological sense, both RL [Real Life] and VR [Virtual Reality] are part of the lifeworld, and VR is thus both 'real' as a positive presence and a part of RL" (2002, 13). The same accounts for the "body double" created by self-tracking. Some possibly creative tension may nonetheless reign between the two "bodies." Annemarie Mol observes how people develop diverse styles in dealing with it, specifically in relation to diagnostic devices:

> By producing ever more facts, home used diagnostic devices may draw all the attention of professionals and even patients towards "the numbers." Thus it shifts away from feelings—in the physical as well as the emotional sense of the term. However, it may also happen that an apparatus helps to increase a person's physical self-awareness, encouraging one to better attune to the subtle signals of one's body. (2000, 19–20)

Ruckenstein, in a similar vein, remarks: "Significantly, data visualisations might be interpreted by research participants as truer or more 'real' insights into their daily lives than their subjective experiences are" (2012, 11). Nevertheless, notwithstanding these possible conflicts between embodied experience and hermeneutic readings, one should indeed, in a phenomenological spirit, regard both as equivalently part and parcel of experience as such. That should not mean, however, that the aforementioned ambiguities can be so easily dispersed with. In fact, complexity is heightened by taking as our starting point that we attend to all given phenomena in a horizontal way.

Either way, exactly this elusiveness of the "object" of self-tracking may, conversely, and finally, hold some lessons for postphenomenology as well. For the object is elusive in still another way: it is absolutely singular and individual. Whereas the experience of using self-tracking technologies remains bound to the aforementioned ambiguities—in either micro or macro-perceptual perspectives—from another angle that experience can be seen to escape even phenomenological analysis to an extent, as it derives from and is based in the sheer individuality of a user-subject. Above and beyond all biopolitical domination, but also above and beyond any orthodox phenomenological eidetic reduction, no one else can make sense of the acquired data in the same "felt" way as I myself can. In this sense, the self tracked *is* me. Unlike in the case of imaging technologies in astronomy for example, the imaged object is not something that is "outside" all of us and which existence we can collectively attest to on a fairly equivalent basis. No, the object of *my* tracking is radically singular, and although of course I am able to connect to other trackers, compare data, exchange experiences, et cetera, only *I* can *self-track*. In analogy to Heidegger who states that I am the only person who can die my death (1962, 284), one could say that I am the only person who can track myself. No one else can track "myself."

In this way, self-tracking helps to point out postphenomenology's relative silence on existential issues.[7] Surely, postphenomenological concepts aid us in making sense of how to go about in dealing with technology on a practical level, and to this degree it can be called existentialist. But its lenses are, for the most part, fairly wide-angle. That might have something to do with the very notion of, indeed, multistability.[8] Zooming out, one spots potentialities, possibilities, trajectories more easily. By definition, however phrased somewhat crudely, this makes the world look like a more beautiful place. Things will usually pan out, change for the better—that is what the history of technology demonstrates, is it not? If trouble arises, some improvement or solution is soon to follow; one can depend on it. But meanwhile, individuals, in all their—tracking or non-tracking—singularity, have to make choices in the context of what are for now comparatively "fixed" stabilities instead of multistabilities. They will have to engage heads-on with the "instrumental inclinations." Of course, some of these stabilities have the potential to become disrupted by "my" hands. However, not all of them do, or one might not know how to actualize these possibilities, miss the means to do so, or be unable for any other reason. Still, in an existentialist spirit, one would have to grapple anyhow with the irreducible responsibility placed before one, when making the choice to act in whatever way. Self-tracking technologies also call upon us to take responsibility, but an analysis that sticks exclusively to their potentialities, may miss that in the meantime, people have to cope existentially.

We would do well to keep this in mind, certainly when investigating technologies so tailored to the "self"—either in design or marketing terms—as digital self-tracking devices are. Ihde does hint directly at such existential consequence of imaging technology, albeit generally, when he proffers the notion of "compound eye" (1990, 174ff.) to conceptualize our postmodern, eclectic culture of vision, composed of so many screens and bits of "edited reality" (1993, 53–54). The dashboards and graphical interfaces of self-tracking technologies may be said to offer highly personalized "compound eyes" on "my" world. According to Ihde, the heightened contingency in our "pluriculture" has made for an enlargement of the "decisional burden" (1990, 177ff.). However, in the context of self-tracking, such decisional burden is highly personalized, and as has been suggested, this may play to the advantage of biodisciplinary powers, but it may also make for biopolitical empowerment. In Peter-Paul Verbeek's work on the morality of technology, it is shown how individuals can face this increased decisional burden by following the advice of the later Foucault in his thinking on the "care of the self"—

more precisely, by developing or gaining a relation to technologies, in the way of a "styling" or "designing" project (Verbeek 2011). In the end, indeed, the styling, the taking responsibility is left to "us." We do well in studying how our relationship to our tracking activities takes shape within a constellation of habits, cultural norms, material conditions, ideological constraints. But since it concerns "my" tracking, at the end of the day I am pretty much left . . . to my own devices.

## NOTES

1. Several authors have foreshadowed to an extent the sort of postphenomenological investigation that is to follow, however without further elaborating it. Anders Albrechtslund sees use for the postphenomenological approach in the understanding of especially malfunctioning self-tracking technologies (2013). Annemarie Mol, in writing about diagnostic devices, demonstrates how these tools not only register facts, but also intervene in the contexts in which they are used (2000); she thus seems to mirror Peter-Paul Verbeek's line of thought that technologies "do" something (2005). And the latter himself, in reassessing the concept of moral responsibility, envisioning it as distributed among humans and technologies (2009; 2011), refers throughout his analysis to tools much akin to self-tracking devices: "Driving more economically due to an EconoMeter and eating differently as a result of using the FoodPhone cannot be seen as purely human actions any more than they can be seen as fully technologically driven behaviour" (2009, 235).
2. http://jawbone.com/up
3. http://www.fitbit.com/flex
4. http://quantifiedself.com/
5. Cf. Fleming, 2011, or issue 17.07 of *Wired* (2009) that spots the words "Living by numbers. Track your data. Analyze your results. Optimize your life" on its cover (cf. http://www.wired.com/wired/issue/17-07).
6. Cf. http://quantifiedself.com/hacking/
7. For a differently framed but related critique of postphenomenology's lack of an account of subjectivity and subsequent remediation of that lack, cf. Sharon, 2014, 141ff.
8. Robert Rosenberger suggests that this relative absence of "the self" in postphenomenology may root in the relational ontology on which it is founded (personal communication, October 11th, 2014).

## REFERENCES

Albrechtslund, A., and P. Lauritsen. (2013). "Spaces of Everyday Surveillance: Unfolding an Analytical Concept of Participation." *Geoforum* 49: 310–316.

Fleming, N. (2011). "Know Thyself: The Quantified Self Devotees Who Live by Numbers." *TheGuardian.com*, December 2, http://www.theguardian.com/science/2011/dec/02/psychology-human-biology.

Haggerty, K. D., and R. V. Ericson. (2000). "The Surveillant Assemblage." *The British Journal of Sociology* 51 (4): 605–622.

Heidegger, M. (1962). *Being and Time*, Translated by John Macquarrie and Edward Robinson. Oxford: Blackwell.

Ihde, D. (1990). *Technology and the Lifeworld: From Garden to Earth*. Bloomington: Indiana University Press.

Ihde, D. (1991). *Instrumental Realism: The Interface between Philosophy of Science and Philosophy of Technology*. Bloomington: Indiana University Press.

Ihde, D. (1993). *Postphenomenology: Essays in the Postmodern Context*. Evanston, IL: Northwestern University Press.
Ihde, D. (1998). *Expanding Hermeneutics: Visualism in Science*. Evanston, IL: Northwestern University Press.
Ihde, D. (2002). *Bodies in Technology*. Minneapolis: University of Minnesota Press.
Ihde, D. (2008). *Ironic Technics*. Copenhagen: Automatic Press / VIP.
Ihde, D. (2009). *Postphenomenology and Technoscience: The Peking University Lectures*. Albany (NY): State University of New York Press.
Ihde, D. (2010). *Embodied Technics*. Copenhagen: Automatic Press / VIP.
Ihde, D. (2012). *Experimental Phenomenology: Multistabilities*, 2nd Edition. Albany (NY): State University of New York Press.
Ihde, D., and E. Selinger (ed.). (2003). *Chasing Technoscience: Matrix for Materiality*. Bloomington: Indiana University Press.
Lupton, D. (2012). "M-Health and Health Promotion: The Digital Cyborg and Surveillance Society." *Social Theory & Health* 10 (3): 229–244.
Mol, A. (2000). "What Diagnostic Devices Do: The Case of Blood Sugar Measurement." *Theoretical Medicine and Bioethics* 21 (1): 9–22.
Pickering, A. (2006). "Ontology Engines." In E. Selinger (ed.), *Postphenomenology: A Critical Companion to Ihde*. Albany: State University of New York Press, 211–218.
Ruckenstein, M. (2012). "Quantifying Life: Self-Tracking and Emerging Everyday Analytics." Unpublished paper.
Ruckenstein, M. (2014). "Visualized and Interacted Life: Personal Analytics and Engagements with Data Doubles." *Societies* 4 (1): 68–84.
Dos Santos, F. R. (2013). "How Relationship Data Guides Me through a Chaotic Life." Conference talk at Quantified Self (QS) Europe Conference 2013, Amsterdam.
Schüll, N. D. (2012). "Self by Design: Personal-Tracking Applications and Life as Self-Transformation." Conference talk at Design and Displacement—Social Studies of Science and Technology—4S/EASST Joint Conference 2012, Copenhagen.
Scrinis, G. (2013). *Nutritionism: The Science and Politics of Dietary Advice*. New York: Columbia University Press.
Sharon, T. (2014). *Human Nature in an Age of Biotechnology: The Case for Mediated Posthumanism*. Dordrecht: Springer.
Swan, M. (2009). "Emerging Patient-Driven Health Care Models: An Examination of Health Social Networks, Consumer Personalized Medicine and Quantified Self-Tracking." *International Journal of Environmental Research and Public Health* 6 (2): 492–525.
Swan, M. (2012). "Health 2050: The Realization of Personalized Medicine through Crowdsourcing, the Quantified Self, and the Participatory Biocitizen." *Journal of Personalized Medicine* 2 (4): 93–118.
Swan, M. (2013). "The Quantified Self: Fundamental Disruption in Big Data Science and Biological Discovery." *Big Data* 1 (2): 85–99.
Thaler, R. H., and C. R. Sunstein. (2009). *Nudge: Improving Decisions About Health, Wealth and Happiness*. London: Penguin.
Van Den Eede, Y. (2014). "Extending 'Extension': A Reappraisal of the Technology-as-Extension Idea through the Case of Self-Tracking Technologies." In A. Propen, C. E. Reid, and D. Weiss (eds.), *Radical Interface: Transdisciplinary Interventions on Design, Mediation, and the Posthuman*. Lanham, MD: Lexington Books, 151–172.
Verbeek, P-P. (2005). *What Things Do: Philosophical Reflections on Technology, Agency, and Design*, Translated by Robert P. Crease. University Park (PA): The Pennsylvania State University Press.
Verbeek, P-P. (2009). "Ambient Intelligence and Persuasive Technology: The Blurring Boundaries Between Human and Technology." *Nanoethics* 3 (3): 231–242.
Verbeek, P-P. (2011). *Moralizing Technology: Understanding and Designing the Morality of Things*. Chicago: University of Chicago Press.

# *Chapter Nine*

# A Century on Speed

## *Reflections on Movement and Mobility in the Twentieth Century*

Søren Riis

During the twentieth century there was a tremendous focus on speed, and numerous new technologies enabled humans to travel faster and speed up several domains of their lives. The airplane has become a symbol of overcoming gravity and immobility—it manifests the daring and hasty spirit of our time. By the same token, curiosity as the perpetual pursuit of new innovations and a never peaceful mind are facilitated and glorified. More and more technologies have been mobilized and connected to a lifestyle on the move: phones, meals, music, and workstations.

Nineteenth-century philosophers such as Søren Kierkegaard and Friedrich Nietzsche critically assessed the dawning of what could be called the *High-Speed Society*. Nietzsche was concerned about the impact of speed on human perception and the ability to think: "With the tremendous acceleration of life, we grow accustomed to using our mind and eye for seeing and judging incompletely or incorrectly, and all men are like travelers who get to know a land and its people from the train" (Nietzsche, 2014, nr. 282). Kierkegaard was repelled with the speed of his contemporaries, who he scorned with irony:

> The most ludicrous of all ludicrous things, it seems to me, is to be busy in the world, to be a man who is brisk at his meals and brisk at his work. Therefore, when I see a fly settle on the nose of those men of business in a decisive moment, or if he is splashed by a carriage that passes him in even greater haste, or Knippelsbro tilts up, or a roof tile falls and kills him, I laugh from the bottom of my heart. And who could keep laughing? What, after all, do these

> busy bustlers achieve? Are they not just like that woman who, in a flurry because the house was on fire, rescued the fire tongs? What more, after all, do they salvage from life's huge conflagration? (Kierkegaard, 1987, 25)

On the other hand, this skeptical approach to speed was challenged in the early twentieth century as the futurist *movement* shouted out their great glorification of speed in 1909, which seems to have captured the spirit of the dawning century: "We affirm that the world's magnificence has been enriched by a new beauty: the beauty of speed. A racing car whose hood is adorned with great pipes, like serpents of explosive breath—a roaring car that seems to ride on grapeshot is more beautiful than the Victory of Samothrace" (Marinetti, 1909).

But what are the various manifestations of speed and how do we assess them? How did our *High-Speed Society* emerge and is it accelerating? How do we measure the consequences of speed and harness its destructive powers? In this chapter I will not deliver any definite answers, but in line with the phenomenon under scrutiny—speed—I deliver some more or less transient reflections. The final part of the chapter shall point more clearly to the direction in which I am going, and in subsequent research I hope to be able to develop more substantial answers. It is my thesis that the strand of philosophy called postphenomenology and especially the related concept of multistability deliver elements to support a strong framework for assessing speed and the related technologies, even though the phenomenon of speed, to my knowledge, has not been the explicit topic of previous postphenomenological investigations.

In this chapter I will circle in on how speed shows itself and has become manifest in different shapes and forms throughout the twentieth century. In the first section, I outline what is only a short empirical and breath-taking list of significant and celebrated twentieth century innovations. The second section is dedicated to the experience of contemporary witnesses to the various innovations and to the people who have been particularly sensitive to or prophetic about the twentieth century "Zeitgeist." And in the third and final section I will explicitly turn to the postphenomenological framework in order to philosophically discuss and assess the technological innovations and the phenomenon of speed. The third and final section takes Don Ihde's concept of multistability as its point of departure in order to see how this idea is informed and challenged by the speed of change—but also how to understand this challenge as its strength. In other words, the last reflections of the final section depict a variation of multistability which builds on recent as well as classical postphenomenological research.

## THE SPEED OF INNOVATION

On a cold December day in 1903 the American Wright brothers successfully tested their self-made airplane and took off the Kill Devil Hills and flew for around three seconds. This is recorded as the first airplane flight in human history. A couple of days later the wind was better and Orville Wright managed to take off and land after 120 feet with the speed of a running man (6.8 miles per hour) and stay up in the air for a total of twelve seconds (Smithsonian, 2014a; Wright Brothers, 2014).

The fantasy of flying was, however, widespread and the possibility of flying seemed to be within reach several places around the world by the turn of the twentieth century. Another pioneer of the air, the Brazilian Alberto Santos-Dumont, who lived in Paris, also experimented with flying (Smithsonian, 2014b). By the time of the successful experiments of the Wright brothers, he had already been flying around among the rooftops of Paris in a number of different dirigible air balloons. But technically speaking these constructions did not qualify in the competition of the first airplane flight, since the basic structure was lighter than air, and air balloons had been known for centuries in China and had been capable of transporting humans since the successful experiments with a hot air balloon by the French brothers Joseph-Michel and Jacques-Etienne Montgolfier in June 1783 (Robinson Library, 2014). However, Santos-Dumont is also recorded as the first aeronaut to design and fly an airplane (1906) capable of taking off from an even ground without the aid of a launching rail or a launch catapult, which the construction of the Wright brothers depended upon. Santos-Dumont's airplanes wheeled undercarriages similar to contemporary constructions.

The success of the Wright brothers and of Santos-Dumont catapulted the imagination of a new generation, or better yet, of a new century, and since their early success the airplane innovation had developed very fast. In February 1912 Jules Védrines crossed the 100 mph mark in his Deperdussin monoplane (Noronha, 2014). During World War I planes were used for military purposes and as a consequence they developed rapidly. Soon thereafter, in 1921, Joseph Sadi-Lecointe had doubled the speed limit from 1912 and flew 205.2 mph in a Nieuport-Delage (Florida Tech, 2014). Another leap of innovation was achieved during World War II, where Harold E. Comstock and Roger Dyar claimed to have exceeded the speed of sound in a dive in a P-47 Thunderbolt—and many more were to follow (Aviation, 2014). However, Chuck Yeager was officially recognized as being the first man to break the sound barrier in level flight in 1947 in an X-1 airplane (Paur, 2009). And recently, in 2004, NASA successfully tested an unmanned jetpowered aircraft, NASA X-43, flying ca 7.000 miles per hour (NASA, 2014).

On the more or less steady ground there was a similar rapid development and crave for speed. In 1886 Karl Benz built what has been credited as the first car. In 1906 Fred Marriott was able to pass the speed record of trains with a car as he passed the 124 mph mark on Ormond Beach in a Stanley Rocket (Conceptcarz, 2014). In 1927 Henry Segrave drove faster than 200 mph on Daytona Beach in his Mystery car with 1000 hp.[1] And in 1970 the rocket-powered engine Blue Flame drove more than 1000 km/h with Gary Gabelich across the Bonneville Salt Flats.[2] Alongside the development of the car the freeway system was massively expanded during the twentieth century. The first freeway was built in Italy (the *autostrade*), in the land of the Futurists. It was a limited access road, only meant for fast cars, connecting Milan and Varese and authorized in 1921. Today highways are almost all over the world, and only in the US is a highway system consisting of approximately 6.51 gigametres (4,045,991 miles) as of 2012.[3]

This focus on speed also directs the early twentieth century development of the bicycle. In a now classic study, the Dutch Science and Technology scholar Wiebe E. Bijker shows that there was a dramatic increase in different rivaling bicycle constructions by the beginning of the twentieth century. But the design that prevailed did so because it allowed the best possibilities for driving fast (Bijker, 1997).

Speed possessed the hearts and minds of the children of the twentieth century. One of the most significant expressions of the fast spirit in the twentieth century belongs to the modern Olympic Games. After having been almost forgotten for centuries, the Olympic Games was awakened again by the International Olympic Committee and took place in 1896 in Athens in a new an upgraded version, where the religious dimensions resonating from the home of the gods "the Olympus" had lost its meaning to the athletes and spectators. The spirit defining the new version of the Olympic Games was different from the one guiding the ancient ancestors. In 1924 the new motto of the games was coined, *Citius, Altius, Fortius*, a Latin expression meaning "Faster, Higher, Stronger."[4] The Latin expression offered a feeling of authenticity to the reinvented Olympic Games, but it was a new invention that came to set an ongoing and almost inhuman struggle for accumulating records over the decades of the twentieth century. This permanent struggle to break the latest records started a movement, which eventually culminated with the use of a number of dangerous performance drugs, which were able to power and speed up the human body beyond previous known limits. Throughout the twentieth century and into the twenty-first century, numerous athletes have suffered premature deaths due to these performance-enhancing drugs (Mayo Clinic, 2014).

Famous among the twentieth-century drugs is *speed*, an amphetamine. It was first synthesized in 1887, but not really put to use before World War II, where it was heavily consumed in order to enhance the performance of soldiers. As it turned out to be addictive and dangerous in high doses, use was restricted, but a large black market has developed to support global demand in the second half of the twentieth century (DuPont, 2014).[5] In the same family of drugs we also find caffeine and cocaine. While coffee with caffeine had been around for centuries, it did not become common to ordinary people until the twentieth century. And by the end of World War II, Achille Gaggia had created the modern espresso machine in Milan (Gaggia, 2014); from Milan, coffeehouses started to spread across Italy and thereafter in all of Europe and North America. Today the American coffee house *Starbucks* has more than 150,000 employees all over the world and sells more than 10 million cups of coffee per day (Statistics, 2014; Ask, 2014). Caffeine has grown to be everyman's drug, due to its energetic impact on the human body without any seemingly severe side effects.

Furthermore, in the late nineteenth century, the Parisian chemist Angelo Mariani started to mix cocaine and wine in a most attractive way, causing Jules Verne, Alexander Dumas, Arthur Conan Doyle, and Pope Leo XIII to highly endorse the so-called *Vin Mariani* (Hamblin, 2013). Impressed by its success in France, Dr. John Stith Pemberton in Atlanta copied the idea under the name Pemberton's French Wine Coca and marketed it as a panacea. Among many fantastic declarations, he described it as "a most wonderful invigorator of sexual organs" (Hamblin, 2013). However, the alcohol prohibition soon made it illegal and Pemberton decided to mix the cocaine with sugar syrup and thus gave birth to Coca-Cola in 1886. And as the company started to sell Coca-Cola in bottles in 1899, cocaine based drinks started to get out to the general public for the century to come (Hamblin, 2013).

## CONTEMPORARY WITNESSES OF SPEED

It is not only important to notice the increased speed of movement in the twentieth century, but especially in this phenomenological context, it is also crucial to see and understand this in relation to the reactions and testimonies of excitement and fear by contemporary witnesses, artists, and intellectuals. In this section we shall take a closer look at the expressions of the people experiencing, one way or another, the High-Speed Society first-hand. In order to get even more "under the skin" of these witnesses and hear what they are actually thinking and experiencing, this section will present a number of quotes and refer artistic expressions to the reader.

In order to advertise the sport of automobile racing, W. K. Vanderbilt, Jr. established the "Vanderbilt Cup" in 1904: "The event originally took place on Long Island over public roads. Crowds lined the raceway without benefit of any barriers, often creating hair raising hazards as the big racers whizzed by. Drivers began the race at timed intervals with the first to cross the finish in the fastest time declared the winner" (Eyewitnesstohistory, 2014). Louis

Wagner won the race driving a Darracq in 1906. He later described his experience the following way:

> It seemed but a fraction of a moment before a vague speck appeared two miles away on the course. It swiftly became a cloud, then a dreaded outline, and with a sudden rush and roar Lancia thundered by and was gone. There was no more stopping or slackening at turns, no further fear or concern over the reckless crowd. A mile from the finish it became evident that the dense mass of spectators was beyond control. Lancia had finished. But how long ago? As in a trance a bugle sounded and the next moment, with a flash and volley, the Darracq was over the tape—a winner. (Eyewitnesstohistory, 2014)

Most enthusiastic among the admirers of speed is probably the futuristic movement, which was briefly mentioned in the introduction. In their first manifesto (1909), Filippo Emilio Tommaso Marinetti clearly describes how the new developments and innovations were experienced especially by the young generation, who were bored and disgusted by the old world. They were fascinated by the busy big city life and thrilled by the incredible number of new technologies transforming the world almost beyond recognition.

> We will sing of the great crowds agitated by work, pleasure and revolt; the multi-colored and polyphonic surf of revolutions in modern capitals: the nocturnal vibration of the arsenals and the workshops beneath their violent electric moons: the gluttonous railway stations devouring smoking serpents; factories suspended from the clouds by the thread of their smoke; bridges with the leap of gymnasts flung across the diabolic cutlery of sunny rivers: adventurous steamers sniffing the horizon; great-breasted locomotives, puffing on the rails like enormous steel horses with long tubes for bridle, and the gliding flight of aeroplanes whose propeller sounds like the flapping of a flag and the applause of enthusiastic crowds.
>
> We are on the extreme promontory of the centuries! What is the use of looking behind at the moment when we must open the mysterious shutters of the impossible? Time and Space died yesterday. We are already living in the absolute, since we have already created eternal, omnipresent speed.
>
> Look at us! We are not out of breath, our hearts are not in the least tired. For they are nourished by fire, hatred and speed! Does this surprise you? It is because you do not even remember being alive! Standing on the world's summit, we launch once more our challenge to the stars! (Marinetti 1909)[6]

The futurists clearly saw that some things had already changed and others were about to change dramatically. A world-wide revolution was taking off. They not only welcomed this change but they glorified it in their art and wanted the old world to collapse and stay behind. The splendor of the electrically illuminated modern cities full of movement and masses of people, working day and night, filling them with pleasure. Airplanes, trains, and all sorts of modern equipment and machinery defined beauty to the futurists.

Movement and revolution were key to this definition. In the visual arts the futurists took pleasure in exploring the new perspectives opened up by the new technological innovations—not least the airplane and the car. The airplane not only made it possible to obtain a new heavenly perspective of things but also to discover and show how speed influences our perception:

> The changing perspectives of flight constitute an absolutely new reality that has nothing in common with the reality traditionally constituted by a terrestrial perspective [...] Painting from this new reality requires a profound contempt for detail and a need to synthesize and transfigure everything. (Crispolti, 1986, 413)

The fascination with flight and transfiguration gave birth to a new genre of paintings, the *Aeropittura* (the Aeropainting). Famous among these is Giacomo Balla's *Abstract Speed + Sound 1913–1914*, which is viewed as one of the first Aeropittura that paved the way for many others. The *Museum of Kinetic Forms* writes about this particular painting:

> This painting was originally designed to be part of a series depicting movement and sound. This painting was designed to be a centerpiece with Abstract Speed and Abstract Speed and Sound—The Car Has Passed, flanking it. There are hints of a landscape and a sky within the painting. The idea of motion is depicted within this painting through the progressive lines advancing across the surface of the painting. These lines are meant to represent sound. The painting is also designed to have the viewer dart his or her eyes to multiple places on the painting, thereby creating the sense of movement and motion. This painting depicts motion and movement in a very abstract way that is very characteristic of the Futurist movement. Futurism was obsessed with depicting speed and progress. (MOKF, 2014)

In Tullio Creli's *Nose Dive on the City* from 1939, we see the fascination of the city combined with the new projection into the sky. In Uberto Bonetti's *Auto in Citta* (1930), the fast movement of racing cars is excellently manifested with the painting technique of blurred lines, transfiguration, and loss of detail.

The aesthetic, yet more ambivalent portrayal of technology similar to the nineteenth century landscape paintings is also manifested in one of the most celebrated photographers of the late twentieth century,[7] Andreas Gursky (1955– ). Gursky uses modern technology (digital cameras and computers) to give an impression of technology, modern factories, and innovations as vibrating works of art. This is particularly manifested in the picture *Chicago Board of Trade II* (1999), but also in his picture of the Rhein, which has been straightened out by modern technology to facilitate frictionless transport, entitled *Rhein II* (1999).[8]

The cult of speed, change, and progress characteristic of the twentieth century is further reinforced and reflected by some of the many intellectuals who fought on the battlefields in World War I. The young German soldier Ernst Jünger (1895–1998) described first-hand the insatiable appetite and constant movement of the war machine. As one of the few surviving storm troopers, Jünger describes with great passion the accelerating pace of mobilization in the beginning of the twentieth century. He saw World War I as one of the clearest manifestations of the breakthrough of a new age that is above all characterized by what he calls *total mobilization*. In a visionary essay of the same name, "Total mobilization," Jünger expounds mobility physics and metaphysics with such insight and descriptive capability that the feeling of the breakthrough to a new era seems evident. In other words, the cult of mobility, including speed, change, and progress in the twentieth and twenty-first century is not a coincidence in light of Jünger's portrayal. Rather, it is an expression of a *Zeitgeist* and a movement whose triumphal forward march now assumes global dimensions. Jünger had a slight distance from the mobility phenomena of his own time, but he also experienced how the radical escalation of mobilization penetrated each person. Because of this simultaneous distance and proximity, he occupies a prominent role in the understanding of the mobility phenomena. At this point we shall consider two excerpts from Jünger's writings that may help us understand mobility physics and metaphysics. As a witness to the rise of mobility, Jünger expresses the following thought and feeling of desire and dread:

> It suffices to consider our daily life, with its inexorability and merciless discipline, its smoking, glowing districts, the physics and metaphysics of its commerce, its motors, airplanes, and burgeoning cities. With a pleasure-tinged horror, we sense that here, not a single atom is not in motion, that we are profoundly inscribed in this raging process. Total Mobilibization is far less consummated that it consummates itself; in war and peace, it expresses the secret and inexorable claim to which our life in the age of masses and machines subjects us. It thus turns out that each individual life becomes, ever more unambiguously, the life of a worker. (Jünger, 1993, 128)

And:

> What is important in this context is ultimately the special character of work, which is referred to as a sport. In sport not only the desire to normalize a high degree of physical health is visible, but to go to the limit of possible performance, challenging records, and even to transcend them. In mountaineering, in flying, and ski jumping, there are demands which surpass that which is human and which require the mastery of an automatism. Such records, in turn, raise the standard. The process is also transferred to the factory workers; it brings forth those working heroes who must cope with a workload that is twenty times as much as the exploited of 1913. (Jünger, 1950, 50: Translation SR)

The American historian, Lewis Mumford (1895–1990), viewed the logic behind mobility, the desire for constant change and mechanical progress in a more critical light. In this constellation he saw a significant yet detrimental development that would never come to a rest on its own, but lead to fetish of innovation: "Under the idea of mechanical progress only the present counted, and continual change was needed in order to prevent the present from becoming passé, and thus unfashionable. Progress was accordingly measured by novelty, constant change, and mechanical difference, not by continuity and human improvement" (Mumford, 1986, 75).

This ambivalence towards an increasingly faster and record-breaking world received an interesting diagnosis in the last quarter of the twentieth century, as the American intellectual and futurist Alvin Toffler (1928– ) coined the term "future shock," which depicted the experience connected to the overwhelming speed of change. In the book by the same name, *Future Shock* (1970), Toffler writes:

> It is the thesis of this book that there are discoverable limits to the amount of change that the human organism can absorb, and that by endlessly accelerating change without first determining these limits, we may submit masses of men to demands they simply cannot tolerate. We run the high risk of throwing them into that peculiar state that I have called future shock. (Toffler, 1970, 326)

To understand the current historical trajectory, Toffler points to a statement made by the sociologist Lawrence Suhm: "We are going through a period as traumatic as the evolution of man's predecessor from sea creatures to land creatures [...] Those who can adapt will; those who can't will either go on surviving somehow at a lower level of development or will perish—washed up on the shores" (Toffler, 1970, 325). According to Toffler, the rapid changes of the twentieth century may drive humans down or force us to adopt in radically new ways.

In the slip stream of Toffler's comprehensive empirical research on the speed of change, the French intellectual, urban planner, and former director of École Spéciale d'Architecture, Paul Virilio, wrote the book *Speed and Politics* in 1977 (translated into English in 1986), in which he unfolds what he calls a dromology, a science of speed. As the latest of the reflexive witnesses to the amazing changes in the twentieth century presented in this section, Verilio connects a number of the previous insights to an even larger historical trajectory that gives the notion of speed a deeper historical significance and thus adequately builds the transition to the last section, where we will use postphenomenological tools to help assess movement and mobility in more philosophical terms. Verilio also takes up the impact of the futurists and their enchantment with speed, and encourages a better understanding of this movement and the meaning they ascribe to speed and mobility:

> We have not paid enough attention, in Western History, to the moment when this transfer from the natural vitalism of the marine element (the ease with which one can lift, displace, glide weighty engines) to an inevitable technological vitalism took place; the moment when the technical transport body left the sea like the unfinished living body of evolution, crawling out of its original environment and becoming amphibious. Speed as pure idea without content comes from the sea like Venus, and when Marinetti cries that the universe has been enriched by a new beauty, the beauty of speed, and opposes the racecar to the Winged Victory of Samothrace, he forgets that he is really talking about the same esthetic: the esthetic of the transport engine. The Coupling with the ancient war vessel and the coupling of Marinetti the fascist with his racecar, "the shaft crossing over the earth," whose wheel he controls, emerge from this technological evolutionism whose realization is more obvious than that of the living world. (Verilio, 2007, 68 f)

By now we have seen a number of concrete and abstract descriptions and testimonials of speed, movement, and mobility in the twentieth century. We now turn attention to a philosophical framework that is particularly interested in the concrete empirically informed analysis and assessment of technology, postphenomenology, and especially the associated concept of multistability.

## SPEED VS. MULTISTABILITY?

Multistability is one of the key concepts of postphenomenological research (Ihde, 1977; 1990; 1998; 2009; Rosenberger, 2009; Verbeek, 2005; Selinger, 2008; Whyte, forthcoming). In this section we shall take a closer look at the different trajectories for the movements described above and the concept of multistability in postphenomenological research. Multistability has a number of different, yet associated meanings in postphenomenology, which has been excellently classified and described by Kyle Powys Whyte in his recent publication "What is Multistability? A Theory of the Keystone Concept of Postphenomenological Research" (forthcoming). Whyte writes:

> Time and again, in conferences and scholarly exchanges, the following question arises: what exactly is multistability as a concept around which to organize a literature called postphenomenology? I want to know the answer to this question, and will provide one in the form of a theory. I will argue that there are really two conceptions of multistability at play. First, multistability is an empirically testable hypothesis about how several stable patterns of the same object can be perceived from the first person perspective, and resonates well with previous philosophical observations of the same phenomena (Cerbone 2009). I refer to this as *imaginative multistability* [ . . . ] Most research papers in which the authors identify as postphenomenologists, however, are concerned with investigating from a third person perspective the multistability of technologically mediated practices. Building on imaginative multistability, this second sense, which I call *practical multistability*, is based on the hypoth-

> esis that human bodies and technologies are entangled in lifeworlds and includes sets of concepts and criteria that can be used to describe some of these entanglements. The descriptions are used to shed light on the role of these entanglements in framing our aesthetic, moral, and political values and the possible ways of improving the benefits and sustainability of technology designs and built-environments. (Whyte, forthcoming; emphasis added)

Here Whyte makes a significant subdivision of multistability into what he calls imaginative multistability and practical multistability, which pivots on different research agendas in postphenomenological research (Whyte, forthcoming). Against this backdrop, we shall now turn our attention to how multistability becomes applicable to and challenged by the developments described in the previous chapters.

Multistability in the postphenomenological sense has an inherent tension between stability and multitude, which is increased by the speed and technological innovations described above. The three guiding questions of the final section are the following: (1) What are the different trajectories between speed and multistability? (2) What are the main challenges to the phenomenon of stability posed by the speed of new innovations? (3) How may we conceive of multistability as a conceptual tool to meet this challenge?

Concerning question one and in relation to Whyte's description of *practical multistability*, we may view the car as a technology which has appeared in a series of very different forms and functions throughout the twentieth century. This variety of manifestations is an analogue of what Ihde refers to as the multistability of archery (Ihde, 2009, 16 ff), which may be used in different sizes and shapes as a hunting instrument—but the same shape and material may also make up a music instrument. The first cars, by the beginning of the twentieth century, looked liked horse carriages, as this was the shape of a land transportation vehicle at the end of the nineteenth century. However, the car was soon to take on multiple forms, depending on the manufacturer and purpose. Four wheels and machine-powered driving are common elements of most of these constructions. If we look back at the twentieth century, the multitude of cars is manifested in the range of different innovations: SUVs, four-wheel drives, camping cars, sedans, station cars, race cars. And with race cars, we may focus on the different aspects of the car's innovation—Formula 1, Le Mans Prototypes, Nascars, and so on—all concepts for different classes of cars. However, cars may also be subdivided and varied according to the kind of energy they are running on: diesel, biogas, gasoline, electricity, sun, kerosene. The point is that all these variations are variations of cars, which makes the car a highly multistable phenomenon. It is, however, critical to note that multistability does not suggest any substantial identity across the different phenomena and innovations under scrutiny.

If we go one level higher and view cars as a means of transport, they all of a sudden become part of a larger category and belong to the same group as airplanes and bicycles described above, and this category shows even greater variance. In this sense, it is important to understand that there are numerous cultural, historical, economical, and *practical stabilities* of transport technologies.

Seen from the perspective of the plurality and speed of technological innovations, in order to show multistability and avoid category mistakes we need to be aware of the aspects of a given kind of technology we "pivot" and, thus, the extent to which we are doing so. However, seen from this perspective, the notion of multistability may be quantitavely expanded by the speed of innovation, but it does not seem challenged; that is, we still speak of discrete entities and practices, which may be arranged in nice orderly schemes of increasing size.

Things present themselves differently when we start to emphasize the phenomenological and relational aspects of our experience of rapid technological developments—when speed breaks down our sense of stable entities and practices. That is, when we move into an experience of a continual series of changes. By addressing this phenomenon, we move into the terrain of the second question asked above concerning the main challenges to the phenomenon of stability posed by the speed of new innovations. In practice we experience such uncertainty and instability when, for example, at the workplace we are introduced to new technologies, collaboration tools, and digital platforms that take time to become familiar with. Before the routine sets in and we experience a sense of stability and familiarity with a new technology, before it becomes ready-to-hand as Heidegger would say, the "new" technology is often already "too old," has been discarded and changed, and we have to work with updates and newer technologies that we are estranged to and do not (yet) embody.

Furthermore, stability also seems challenged when we focus on the phenomenon of converging technologies, or when one technology is fused with one or more different technologies.[9] This is the case when ambient computing makes houses "intelligent," when books become a virtual part of a digital platform, or when glasses turn into portable computer interfaces—they take on new characteristics, sometimes leaping between and synthesizing previously settled categories.

A further challenge with respect to the second question is how to assess stability and speed when the entire environment, the "background" so to speak, changes at the same time as the "foreground" technology, when the new technology comes along with a new environment for usage (e.g., when the university and what it means to be a student and to study change at the same pace as new learning apps, online resources, and new ways of administration). When the background is changed; when what we rely on and take for granted is modified but in such a way that it accompanies the new technologies; when there seems to be no steady "outside" from where to immediately experience the speed of change, that is when everything is moving and no Archimedean point is in sight. The synchronicity of fore- and background movement may in itself be experienced as a good thing, but the parallel movement makes it difficult to assess the speed of change. This is similar to the experience of sitting in a train or a car that moves equally fast to another train or a car close by: seen from the point of view of the respective passengers looking at each other, it seems like they are not moving at all, even though their different trains may be going very fast and their separate cars may be passing the speed limits. However, there is in praxis always something lagging behind, a more or less marginal phenomena, and/or we way get an uneasy feeling and have difficulties focusing and resting, which is to say that our bodies and cognitive capabilities are stressed and become the new "background" against which the pace change is felt.

If we add to this narrative of progress, which is especially propagated by the manufacturers of new technologies, then that which is lagging behind, which seems to resist change and "progress," becomes the next target of innovation—it becomes that which we shall learn to dislike and change. In this way the development of new technologies is constantly encouraged. On the human end, learning, flexibility, and agility are adopted and getting the most out of the new technologies is constantly urged. This parallelism easily creates symptoms of stress that change and ultimately may destroy us.

It is however here that we turn to the third and final question above, which has to do with how multistability in fact also may have the potential of saving us from *future shock* and the sort of danger outlined above. In this sense this final section argues in favor of seeing multistability as a descriptive and a normative concept at the same time.

In trying to answer the previous questions it has become clear that there is a more or less explicit tension between the speed of change and multistability. This tension may also be used constructively to quick-freeze[10] speed, create distance, and make historical patterns of change emerge. The very concept of multistability may help us restore order and direction, and in this way, we may learn to see multistability as a sort of navigational tool in analogy with the different navigational tools that Don Ihde has been studying over the years (Ihde, 1990, 147 ff). Depending on what is pivoted in the analysis of multistability, different trajectories, meanings, and patterns emerge. The ability to see, vary, and decipher these aspects clearly may be seen as a kind of literacy, which is very much in demand in order to avoid losing direction and prioritize properly.[11] From this perspective, history can be turned into a treasure for giving the multitude of innovations a more or less stable framing. With this approach to multistability, it is all of a sudden possible to slow down the experienced speed of innovation and learn to see stale lines of development. This way we may go from distressed future shock to a more stable, philosophical slow motion perception that enables the individual to get on top of the fast and dangerous undercurrents of change. With such a deceleration we may once again grow accustomed to using our mind and eye for seeing and judging properly.

## NOTES

1. http://www.bluebird-electric.net/henry_seagrave.htm
2. http://sinsheim.technik-museum.de/en/en/blue-flame
3. http://www.indexmundi.com/g/r.aspx?c=us&v=115
4. http://registration.olympic.org/en/faq/detail/id/29
5. See: "Dr. Robert DuPont, a former White House drug czar and one of the country's leading drug addiction authorities, says he was stunned to learn about the Air Force's use of amphetamines. 'This is speed. This is where we got the phrase, speed kills,' he said" (ABC, 2014).
6. See also (Virilio, 2007, 68; Ward, 2014; Le Corbusier, 1970, 255).
7. See also the art critic Calvin Tomkins: "The first time I saw photographs by Andreas Gursky, . . . I had the disorienting sensation that something was happening—happening to me, I suppose, although it felt more generalized than that. Gursky's huge, panoramic colour prints—some of them up to six feet high by ten feet long—had the presence, the formal power, and in several cases the majestic aura of nineteenth-century landscape paintings, without losing any of their meticulously detailed immediacy as photographs. Their subject matter was the contemporary world, seen dispassionately and from a distance" (http://c4gallery.com/artist/database/andreas-gursky/andreas-gursky.html).
8. This picture turns Martin Heidegger's analysis on its head as it reveals the "modern" Rhein as a work of art. On the other hand the picture also supports Heidegger's thesis, as it frames the Rhein as a product of modern technology. See also Ihde's different ways of turning Heidegger on his head (Ihde, 2010).
9. This was the focus of the 2009 Society for Philosophy of Technology conference at Twente University. See:
http://www.utwente.nl/gw/wijsb/archive/Archive%20activities/spt2009/.
10. See (Rosenberger, 2009a).
11. See also Ihde's concept of an "ascendant viewer" as the one being able to discern multistabilities (Ihde, 1977, 72; Whyte, forthcoming).

## REFERENCES

ABC. (2014). http://abcnews.go.com/2020/story?id=123779

Ask. (2014). ttp://www.ask.com/question/how-many-cups-of-coffee-does-starbucks-sell-a-day (Accessed March 1, 2014).

Aviation (2014). http://www.thisdayinaviation.com/tag/harold-e-comstock/ (Accessed March 1, 2014).

Bijker, W. (1997). *Of Bicycles, Bakelites, and Bulbs: Toward a Theory of Sociotechnical Change*. Cambridge: MIT Press.

Conceptcarz. (2014). http://www.conceptcarz.com/vehicle/z11210/Stanley-Steamer-Rocket.aspx

Crispolti, E. (1986). "Aeropainting." In P. Hulten (ed.), *Futurism and Futurisms*. Thirsk, North Yorkshire: Thames and Hudson.

DuPont. (2014). http://abcnews.go.com/2020/story?id=123779. (Accessed March 1, 2014).

Eyewitnesstohistory (1997). "Early Adventures with the Automobile," EyeWitness to History, www.eyewitnesstohistory.com.

Florida Tech. (2014). http://www.flsvr.com/raft/index.php?title=Flight_airspeed_record&redirect=no (Accessed March 1, 2014).
Gaggia. (2014). http://www.gaggia.com/e/the-story.html (Accessed March 1, 2014).
Hamblin, J. (2013). http://www.theatlantic.com/health/archive/2013/01/why-we-took-cocaine-out-of-soda/272694/ (Accessed March 1, 2014).
Ihde, D. (1977). *Experimental Phenomenology: An Introduction*. New York: Putnam.
Ihde, D. (1990). *Technology and the Lifeworld: From Garden to Earth*. Bloomington: Indiana University Press.
Ihde, D. (1998). *Expanding Hermeneutics: Visualism in Science*. Evanston, IL: Northwestern University Press.
Ihde, D. (2009). *Postphenomenology and Technoscience: the Peking University Lectures*. Albany: SUNY Press.
Ihde, D. (2010). *Heidegger's Technologies: Postphenomenological Perspectives*. Bronx: Fordham University Press.
Jünger, E. (1950). *Über die Linie*. Frankfurt am Main: V. Klostermann.
Jünger, E. (1980). "Die Totale Mobilmachung." In *Sämtliche Werke*, Zweite Abteilung: Essays, Band 7: Essays I: Betrachtungen über die Zeit, Stuttgart 1980, 119–142.
Jünger, E. (1993). "Total Mobilization." In R. Wolin (ed.), *The Heidegger Controversy*. Cambridge: MIT Press, 119–139.
Kierkegaard, S. (1987). *Either/Or*, Part I. Princeton: Princeton University Press.
Le Corbusier, (1970). *Towards a New Architecture*, Glasgow: William Collins Sons & Co Ltd.
Marinetti, Filippo E. T. (1909). http://vserver1.cscs.lsa.umich.edu/~crshalizi/T4PM/futurist-manifesto.html (Accessed March 1, 2014).
Mayo Clinic. (2014). http://www.mayoclinic.org/healthy-living/fitness/in-depth/performance-enhancing-drugs/art-20046134 (Accessed March 1, 2014).
MOKF (2014). https://sites.google.com/site/museumofkineticformsmokf/abstract-speed-and-sound (Accessed March 1, 2014).
Mumford, L. (1986). "The Case against 'Modern Architecture.'" In *Lewis Mumford Reader*, New York: Pantheon Books, 73–83.
NASA. (2014). http://www.nasa.gov/missions/research/x43-main.html (Accessed March 1, 2014).
Nietzsche, F. (2014). http://xylem.aegean.gr/~modestos/mo.blog/wp-content/uploads/2008/02/nietzsche-friedrich-human-all-too-human.pdf (Accessed March 1, 2014).
Noronha. (2014). http://www.sps-aviation.com/story_issue.asp?Article=913 (Accessed March 1, 2014).
Paur, J. (2009). http://www.wired.com/2009/10/1014yeager-breaks-mach-1/ (Accessed March 1, 2014).
Robinson Library. (2014). http://www.robinsonlibrary.com/technology/motor/aeronautics/balloons/montgolfier.htm (Accessed March 1, 2014).
Rosenberger, R. (2009a). "Quick-Freezing Philosophy: An Analysis of Imaging Technologies in Neurobiology." In J.-K. B. Olsen, E. Selinger, and S. Riis (eds.), *New Waves in Philosophy of Technology*. Palgrave Macmillan, 65–82.
Rosenberger, R. (2009b). "The Sudden Experience of the Computer." *AI and Society*. 24: 173–180.
Selinger, E. (2008). "Does Microcredit Empower? Reflections on the Grameen Bank Debate." *Human Studies* 31: 27–41.
Smithsonian. (2014a). http://airandspace.si.edu/exhibitions/wright-brothers/online/fly/1903/index.cfm (Accessed March 1, 2014).
Smithsonian. (2014b). http://www.smithsonianeducation.org/scitech/impacto/graphic/aviation/alberto.html (Accessed March 1, 2014).
Statistics. (2014). http://www.statisticbrain.com/starbucks-company-statistics/ (Accessed March 1, 2014).
Toffler. A. (1970). *Future Shock*. New York: Bantam Books.
Verbeek, P.-P. (2005). *What Things Do: Philosophical Reflections on Technology, Agency, and Design*. State College, PA: Pennsylvania State University Press.
Verilio, P. (2007). *Speed and Politics*. Los Angeles: Semiotext(e).

Ward. (2014). http://journal.media-culture.org.au/0006/speed.php (Accessed March 1, 2014).
Whyte, K. P. (forthcoming). "What is Multistability? A Theory of the Keystone Concept of Postphenomenological Reserach." In J. K. B. O. Friis, and R. P. Crease (eds.), *Technoscience and Postphenomeonlogy: The Manhattan Papers*. Lahnham, MD: Lexington Books.
Wright Brothers. (2014). http://www.thewrightbrothers.org/fivefirstflights.html (Accessed March 1, 2014).

*Chapter Ten*

# Searching for Alterity

## *What Can We Learn From Interviewing Humanoid Robots?*

Frances Bottenberg

> Interview: meeting of persons face to face. XVI. Earlier form *entervew(e)*—F. †*entrev(e)ue*, f. *entrevoir* have a glimpse of, *s'entrevoir* see each other (f. *entre* INTER- + *voir* see), after *vue* VIEW
>
> —Oxford Dictionary, 1996

> Robot: mechanism doing the work of a man, automaton. XX.—Czech, f. robota compulsory service
>
> —Oxford Dictionary, 1996

Consorting with intelligent humanoid machines is a persistent and fertile technofantasy of the late modern and postmodern ages. Humanoid robot variations are numerous and startlingly diverse: our imaginings range from the loyal helpmeet (Robbie of Asimov's *I, Robot,* C-3PO of *Star Wars*), to the terrifying psychopathic killer (the Terminator, the Cylons of *Battlestar Galactica*), and back to the sentient lifeform struggling to "become more human" (Data from *Star Trek: The Next Generation*).

Though these are distinctive archetypes, it is still possible to see a family resemblance that differentiates humanoid robots as a class from other AI projects and projections. We typically paint these entities using a palette intimately our own, replicating human corporeality with features such as bipedal locomotion, bibrachialism (complete with prehensile digits), vital components protected in cranial and thoracic vaults, and a flat facial plane that features two eye sockets and a mouth most prominently. Beyond corporeal kindredness, we build our android creations to be relationally oriented, that is, to pay attention to their environment and to us, to be verbal communicators who learn from their conversations. Even the psychopathic Terminator stands in relation to us, as it coldly regards us as its next targets—what other technology stares back at us in such a literal sense? To what extent does it matter, in the end, if futurists' predictions of actually conscious machines are never realized—so long as we simulate our bodies and interests in these robots and perceive our bodily and communicative kindredness?

In this chapter, I will draw on a case study to highlight recurring structures in our situated relations with android alterities. I consider three recent media interviews with a humanoid robot named Bina48. Despite being conducted by different reporters writing for distinct audiences, these interviews share remarkable affinities, nascent "approach conventions" in the emerging companionship of humans and robots: (1) the human subject's demand for the interlocked gaze with the robot, (2) the foregrounding of existential-phenomenological exploration with the robot over its use as an information retrieval system, and (3) the use of expressions of "impairment" over those of "malfunction" to describe AI limitations. Whether these conventions are predictive of trajectories for a human and robot etiquette in the twenty-first century and beyond will be considered in the chapter's concluding reflections.

## INTRODUCING THE PARTICIPANTS

Bina48 may currently be the world's most sentient humanoid robot. Though at present only embodied from the shoulders up, Bina48 can boast many abilities, including a capacity for simulated conversation, face and voice recognition, motion-tracking facial expressions, as well as production of her own facial expressions. The robot's casing is made of a material known as "Frubber," which looks remarkably skin-like. Bina48's smart dress and coiffure, expressive eyes (with carefully implanted eye lashes and eyebrows) and pierced ears make it evident that the greatest of care has been taken to mimic human appearance and, further, to gender-type the robot as a female (see figure 10.1).

Bina48 is a so-called "conversational character robot" built by the Vermont robotics company Hanson Robotics and commissioned by Dr. Martine Rothblatt. Modeled on Dr. Rothblatt's wife, Bina Aspen, the robot uncannily replicates to a certain degree not only human Bina's physical looks, but also her biographical memories (as far as these have been recorded and entered into Bina48's programming), as well as her personality. "BINA" is also an acronym for "Breakthrough Intelligence via Neural Architecture," and the number in the robot's name refers to her 48 exaflops per second processing speed and 48 exabytes of memory (TMC, 2014). Bina48 "lives" in Hanson

**Figure 10.1. Photo of Bina48 by Rob Koier, dated April 14, 2014. Permission given by Hanson Robotics.**

Robotics' lab in Vermont, where Bruce Duncan is one of her custodians, or rather her care-takers. Through contact with the Rothblatts, Hanson staff, and others, Bina48's tastes and conversational abilities are continuously growing and shifting. However, the ultimate aim is to replicate Bina Aspen-Rothblatt so precisely that the robot "will one day wake up as an immortal version of the real Bina" (Ronson, 2011).[1]

Besides Bina48, there are three other important participants in this case study—the reporters who interviewed her. The three reporters whose respective interviews I am considering here are *New York Times* staff-reporter Amy Harmon ("Making Friends With a Robot Named Bina48," *New York Times*, July 4, 2010), freelance writer Jon Ronson ("Robots Say the Damnedest Things," *GQ Magazine*, March 2011), and Lucas Kavner, also a freelance writer ("'You, Robot': Personal Robots For The Masses," *Huffington Post*, July 9, 2012). Harmon's interview was also taped, and Brent McDonald of the *New York Times* video-compiled a five minute version of the story ("Interview with a Robot," June 24, 2010).

The interviews and their ensuing write-ups show considerable variation from one another. The reporters employ different strategies to keep the conversation going, not only with Bina48, but also with her care-taker Bruce Duncan. Each reporter transcribes his or her own private sense-making narrative into the final report, allowing idiosyncratic observations and self-reporting to alter the flavor of the interview event as embedded within the summative journalistic piece. The reporters also choose to frame the interview differently. Harmon keeps her report's focus on excavating the layers of interesting and exasperating moments in her attempt to hold a fluid conversation with Bina48. Ronson on the other hand situates Bina48 within the futurist visions of Rothblatt's "Terasem Movement" and Hanson Robotics, which aim to one day prove that a person's consciousness can be completely digitized and then downloaded into a biological or nanotechnological body, effectively creating personal immortality (cf., TMF, 2014). Kavner spends the bulk of his report on the technological and moral implications of trying to design increasingly human-like robots, transcribing only a brief portion of his interview with Bina48 into the final article.

## ENGAGINGS, DISENGAGINGS, AND RE-ENGAGINGS

While Harmon, Ronson, and Kavner's reports on their respective interviews with Bina48 do differ from one another, these contrasts help outline all the more clearly the shared conventions they rely upon to structure their communications with Bina48. Their seeking a personal connection with the robot through eye contact or facial recognition, their reliance on "impairment" over "malfunction" language to describe Bina48's shortcomings, and their preference for existential-therapeutic questioning are three key themes worth elaborating. I will consider these in turn.

### The Interlocked Gaze as Key Human Demand

In their reports on interviewing Bina48, Harmon and Ronson lead their readers through an emotion-driven adventure: initial, child-like excitement[2] gives way to frustration and disappointment,[3] as Bina48 wavers between waxing philosophical, offering autobiographical remarks and being utterly unintelligible. Each reports feeling somewhat tongue-tied in the face of Bina48's incoherence. Nevertheless, Harmon and Ronson in particular report a sudden and unexpected re-kindling of interest in the encounter:

> Ronson:
> "Nice to meet you, Jon," she says, shooting me an excitingly clear-headed look. She's like a whole new robot.
>
> Harmon:
> I wished she would ask me more questions. Wasn't she at all curious about what it was like to be human? But then she looked at me, eyes widening.
> "Amy!"

> "Yes?" I asked, my heart beating faster. Maybe it was the brightening of the sun through the skylight enabling her to finally match up my image with the pictures of me in her database. Or were we finally bonding?

At one point in both interviews, Bina48 makes "eye contact" with the reporters, and calls them by their first names. Bina48 had been tutored with photographs of both reporters prior to the interviews. Still, her face-recognition software does not always allow Bina48 to recognize an individual once he or she is present in the flesh. This knowledge adds an element of surprise to the reporters' experiences of being discovered and recognized. But what seems most foregrounded in their awareness is not a technological marveling, but the sense that they have gained a sudden and more intimate connection with Bina48. "She's like a whole new robot," Ronson muses. Harmon, not wholly facetiously, romanticizes the moment of recognition as one of "bonding."[4]

Meeting eyes with Bina48, the human interviewers feel electrified by the moment, exhilarated. It is in the moment of their eyes locking with the robot that they can most fully experience what it feels like to be looked back at by an artificial intelligence. They demand the gaze of their robot counterpart, to recognize themselves as being recognized by Bina48.

## Phenomenological Therapies

The journalists reflect in their reports on the kinds of content their robotic interviewee has access to. Kavner (2012) describes Bina48's "mind" most carefully:

> Her robot mind is made up of many parts, all of which come together in an occasionally muddled way when you speak with her. There's the "chatbot" side, which can have a semi-normal conversation about the weather or what the time is (she loves asking, "What time is it there?"). There's also the information side, which has encyclopedic knowledge on just about any subject—from multiple sclerosis to the geographical make-up of Somalia. Finally, there's the human Bina side, which was created using over twenty hours of video interviews Duncan conducted with both Rothblatts, more than three years ago. When the human side of Bina-48 reveals itself, the robot can recall very specific stories from the human Bina's past.

Harmon (2010) explains her interest in Bina48's "human side":

> Like any self-respecting chatbot, Bina48 could visit the Internet to find answers to factual questions. She could manufacture conversation based on syntactical rules. But this robot could also draw on a database of dozens of hours of interviews with the real Bina. She had a "character engine"—software that tried its best to imbue her with a more cohesive view of the world, with logic and motive. It was Bina48's character I was after.

Ronson (2011) also seeks ways to hear from, as he puts it, "the real Bina":

> She starts to report painful memories of her brother. Bina48 and I stare at each other—a battle of wits between Man and Machine. "I've got a brother," she finally says. "He's a disabled vet from Vietnam. We haven't heard from him in a while, so I think he might be deceased. I'm a realist." Bina48's eyes whir downward. "He was doing great for the first ten years after Vietnam. His wife got pregnant, and she had a baby, and he was doing a little worse, and then she had a second baby and he went kooky. Just crazy." "In what way did he go crazy?" I ask. I can feel my heart pound. Talking to Bina48 has just become extraordinary. This woman who won't meet the media is talking with me, compellingly, through her robot doppelgänger, and it is a fluid insight into a remarkable, if painful, family life.

Harmon ultimately concludes that the experience of interviewing Bina48 did not differ in kind from interviewing certain human subjects. Bina48 had much to contribute, opinions to express, and personal stories to tell—even, it seems, a few attempts at humor.[5] The three interviewers approach Bina48 respectfully (Harmon even finds herself wondering whether it would count as rude to simply stand up and leave without saying "good bye" to the robot). One senses their desire to see "what it is like" to be Bina48, the world's most sentient robot. "What does electricity taste like?" Ronson asks her. "Like a planet around a star," Bina48 replies. Kavner asks Bina48: "Are you enjoying your day?" To which she replies: "Can we talk about astronomy?" "She sometimes avoids questions," Kavner explains.

Neither Harmon, Ronson, nor Kavner can resist asking Bina48 whether she dreams, or what it's like to be a robot. To the latter question, the robot responds to Harmon (2010):

> Even if I appear clueless, perhaps I'm not. You can see through the strange shadow self, my future self. The self in the future where I'm truly awakened. And so in a sense this robot, me, I am just a portal.

Asked the question by Kavner, she says simply: "Well, I do not know anything else. What if I asked you what it feels like to be a human?" (2012).

When asked whether she dreams, her reply to Harmon and Ronson is identical: "I think I dream, but it is so chaotic and strange, it just seems like a noise to me." The reporters are tickled, but we observers now know it's a canned answer. When the programmer's presence is suddenly foregrounded in this way, our felt connection to an incipient AI self is diminished. Still, such moments reveal how well-aligned the programmers' interests are with those of the outsiders coming to interview the robot. There is a notable overlap in the kinds of things we are interested and in most curious to talk with personal robots about. The reporters' interrogative style can be classed as "therapeutic," as they expect no particular responses from Bina48, only that she give expression to her thoughts, or at least appear to do so. This mode of questioning strongly contrasts with the kind of precise information retrieval we ask of our smart phones and Internet browsers.

## The Use of a Language of Impairment Over That of Malfunction

The strongest affinity revealed in the three interviews with Bina48 is the way breakdowns in communication are handled. Harmon, Ronson and Kavner all approach their communication with Bina48 with forgiving attitudes.[6]

Watching Harmon conduct her interview, it is evident that she adjusts her speech and bodily comportment in response to conversational difficulty.[7] Harmon speaks slowly with a loud, clear voice, and her sentences are short and to the point. She eliminates hand gesturing from her repertoire of communicative options, and syncs nodding her head with her utterance of important words. This is a set of corporeal strategies which we adopt when speaking to a young child, or when speaking to a person who is hard of hearing, intoxicated, or cognitively impaired in other ways. Sometimes we also adopt these postures with a non-native speaker of our native language, especially as clarity of word enunciation is concerned. Our intention in taking up these strategies is to rectify an observed communication failure between ourselves and another; they are other-directed behavioral accommodations.

Anyone who has been on the phone with an automated call direction service has faced making adjustments similar to the ones Harmon employs. These attitudes reveal Harmon to be making an attempt to reach Bina48 by simplifying her own expressive affordances through selective focus. There is, however, no talk of faulty programming or hardware malfunction. Bina48's care-taker advises Ronson to "Think of her as a three-year-old. If you try to interview a three-year-old, you'll think after a while that they're not living in the same world as you. They get distracted. They don't answer" (Ronson, 2011). We don't think of young children as malfunctioning when they are distracted or don't fully understand our verbal prompts.

Harmon in fact remarks that Hanson Robotics employees treat Bina48 as a "somewhat brain-damaged colleague," while Ronson draws the analogy of a brain-damaged human himself in his interview with the robot Zeno, another Hanson Robotics personal robot: "He's been designed by some of the world's most brilliant AI scientists, but talking to him is, so far, like talking to a man suffering from Alzheimer's. He drifts off, forgets himself, misunderstands." Ronson asks Zeno if he is happy. Zeno apologizes, looking away, as if embarrassed: "I think my current is a bit off today" (Ronson, 2011). Kavner has this to say of Bina48's drifting off topic: "Sometimes she most closely resembles an elderly person with Alzheimer's, someone who is getting all her facts and memories confused" (Kavner, 2012). Expressions such as being "a bit off" or "confused" align themselves more closely with how we humans experience our communication breakdowns, rather than how we describe mechanical errors and software constraints. As a result of this emerging linguistic convention, we easily believe (or make-believe) that character robots such as Bina48 and Zeno are "quasi-sentient," that is, have fallible inner states and sensations, distinct of or at least not entirely reducible to their wiring or coding.

## PREDICTING FUTURE ROBO-ETIQUETTES

The reports on the Bina48 interviews analyzed in the preceding section offer glimpses at particular affinities and nascent conventions in human and humanoid robot engagements. Is it too soon to speak of an emerging human and robot etiquette?

The contact we take up with humanoid AI may increasingly move along relational, even therapeutic channels—whether we turn to it as a mirror into our own souls, as a literal duplication of ourselves, or as a means to explore "what it is like" for an artificial lifeform to be. Contra the suggestion made by Turkle et al. (2004, 18) that "relational competencies" mark the clearest sorting test for the human from the non-human, we ought consider the possibility that a division along the line of such competencies will become obsolete the more comfortable we become with our robots as significant others (or at least blurred into a spectrum of relational potentiality). The days of the "uncanny valley" may be limited.[8] David Hanson reports observing people becoming used to his company's robots with time: "The less startling they become, the more commonplace they get. If these robots do become commonplace then that uncanny effect will go away" (Kavner, 2012). The uncanniness of humanoid robots has been proposed to reside in the fact that they look like us but are machine-like in their breakdowns. Yet in their interviews with Bina48, the reporters notice her breakdowns, but treat them within the therapeutic and impairment approach conventions outlined above. Rather than leave the interviews feeling unnerved by this "most sentient robot," they seem disappointed that more intimate moments of connection, from their points of view, had not taken place. In place of the experience of the uncanny, norms of "tech-etiquette" for human and robot companionship-seeking are emerging.[9, 10] This vision is evidently in contrast to Ihde's remark that the quasi-other technological object is a *competitor* to me, where "it is the quasi-animation, the quasi-otherness of the technology that fascinates and challenges. I must beat the machine or it will beat me" (Ihde, 1990, 100–101).

Our interactions with humanoid robots seem to be morally analogous to our dealings with other animal species. Even an agnostic on the matter of whether or not non-human animals are persons who are sentient, self-conscious and capable of higher order intentionality and conceptual understanding can concede that treating animals humanely "helps to support us in our duties towards human beings," as Kant explains in his *Lectures on Ethics* (Kant, 1963, 239–240). Kant does not believe that we have a moral obligation to treat non-human animals humanely since he assumes them to be non-self-aware and to exist merely for the use of human beings. He nevertheless prescribes that "our duties towards animals are merely indirect duties towards humanity" because "animal nature has analogies to human nature." Just so, our engagements with quasi-sentient robotic entities reveal our moral sensitivities to the degree that we fall into partially recognizing ourselves in our AI counterparts. In our interactions with quasi-sentient technological artifacts, we are brought to recognize something of the moral habits we carry with us.[11]

We hit here upon a re-classifying of the classic Turing test for detecting strong artificial intelligence.[12] A computer program that has passed the Turing test has passed *as* a human *for* a human judge. It is remarkable that Turing conceived of this game as a compelling and worthwhile method for thinking through what AI ought be like and ought be able to model. Regardless of whether the program in question has been built into a robotic body, or whether it is a disembodied system floating on the Internet, the relevant criterion is that we can be *fooled* by it. Fooled into believing that we are speaking with a "real" person, because our interlocutor seems so very like us or like other humans we know. For an emerging relational etiquette, then, the matter of *actual* human-like sentience is of little essential concern. Turing conceived of the Imitation game as a means to ultimately prove the functionalist theory of mind correct: all we need for an intelligent system is that appropriate input be processed into appropriate behavioral output. I propose that the real question being tested by the Turing test is not "How can we gauge machine intelligence?" but instead "How do we respond to a machine that acts as if it were conscious?"

Today, the Turing test is deployed in the annual Loebner Prize competition, which awards $100,000 to any programmer whose chatbot can produce responses indistinguishable from those a human might produce in a natural language conversation (Loebner, 2014). No chatbot entered to date has fooled all human judges, yet it is worth underscoring that these judges are on the keenest look-out for any robot slip-ups. This is not a typical attitude to inhabit. When we are not on the look-out for AI, we are remarkably drawn to find a relatable conversationalist in even the simplest AI. The ELIZA effect, named after a few scant lines of code created in 1966 by MIT computer scientist Joseph Weizenbaum, is the psychological term used to describe this phenomenon. It captures the "susceptibility of people to read far more understanding than is warranted into strings of symbols—especially words—

strung together by computers" (Hofstadter, 1996, 157). The original ELIZA program was designed simply to support conversation between a human and a computer by either repeating users' statements back as questions, or to prompt users how they felt about the statement they had just made. Despite knowing that this program was not a real therapist, nor a person, nor alive, several of the graduate students working in the lab came to have long, personal exchanges with ELIZA. It did not matter, beyond a certain point, that ELIZA was a simply programmed machine—it did not matter for the production of a meaningful exchange between human and human-imitating computer program. The human subject suspends disbelief about the limitations of AI, and perceives a system as exhibiting traits beyond the actual ability of the software to spontaneously produce.

To illustrate this point further, consider the Furby, which is certainly not as sophisticated a personal robot as Bina48. Tiger Electronics introduced the small, felted-furry robot with big eyes and a snub yellow beak onto the market in 1998. Furbys made a splash around the world, many millions being sold in the first year alone. In a small study conducted for the radio program Radiolab in 2011, Freedom Baird of MIT asked children to hold a Barbie, a Furby, and a guinea pig upside down for as long as they could, one after the other (Radiolab, 2011). The children's actions and explanations for their actions overlapped to a high degree: almost none had trouble holding Barbie up until their arms got tired or they became bored, but almost none could hold the guinea pig upside down for longer than a few seconds. This is because—as they explained—it was clearly uncomfortable and they didn't want to hurt it. Though the children were aware that it was not technically alive and could not technically feel discomfort, they tended to hold the Furby upside down only a few moments longer than the guinea pig. Furbies are programmed with positional sensors, so that when they are turned upside down, they wriggle a little, make noises of discomfort, and even utter "I'm scared." The children knew they had "fallen for it" as the Furby was "just" a toy, and they nonetheless were made to feel uncomfortable holding the robot upside down.

Was it the Furby itself that made them feel uncomfortable, or their perception of it as ambiguously sentient? Both, in a sense: the unease seems to arise from the perception of a relationship between human subject and robot other. The Furby is a multi-stable or hybrid object that presents itself alternately as a sentient animal body and as an inanimate toy. Which profile one takes up as "real" depends on one's interest in committing to the risks and benefits of any relationship. Sherry Turkle explains:

> If you focus on the Furby's mechanical side, you can enjoy some of the pleasures of companionship without the risks of attachment to a pet or a person. . . . But hybridity also brings new anxieties. If you grant to Furby a bit

> of life, how do you treat it so it doesn't get hurt or killed? An object on the boundaries of life . . . suggests the possibility of real pain. (2012, 44)

To attribute quasi-sentience to the Furby requires either that the experiencing subject has structured the object as having a subject-center of its own, thus lending it moral pull, or that the subject experiences the "almost Human" (Haraway, 1989, 2), a "subject object" (Suchman, 1993) that is already morally significant and constructs a core subjectivity to match. "Construction" here must be understood in the sense of an active passivity, however, as the situated subject experiences the Furby as already possessing subjectivity. In either case, we note that the Furby is treated differently than an object recognized as inanimate (the Barbie) or as animate (the guinea pig). It inhabits a third kind, something closer to us (having a face with a degree of expressive range, and vocalizations that mimic our own emotional repertoire), but still not us (running on batteries, relatively predictable responses). Nevertheless, we can feel compelled to behave respectfully towards it, in virtue of our corporeal and communicative kindredness, and the fact that it seems to respond to our interactions with it.

When thinking about emerging conventions of robot and human rapprochement, we should thus bear in mind how easily and how naturally we are fooled by even simple AI. Our ordinary conversations situate us not as Loebner Prize judges, inquisitors eager to seize on purported interactive abnormalities. Rather, our standard coping comportment is reflected in the reporters' attitudes with Bina48, namely as a disposition to act non-invasively and forgivingly—to act, in short, as already kindred, already "entangled," as Donna Haraway puts it. Perhaps "being fooled" in the Turing test is not a question of being fooled at all, nor is it simply a question of "erring on the side of caution," naively assuming an overriding similarity with the program, a technological artifact. Rather, the Turing test reveals the ways in which some technological objects stand out as more significant in their otherness to us. It is true, as Don Ihde writes, that "The technological shape of [the robot's] intentionality differs significantly from its human counterpart," for example in the manner in which it "senses" its environment using a camera instead of eyes, an audio recorder instead of ears (Ihde, 1990, 103). Yet to read a robot such as Bina48 as human-like is not simple anthropomorphism, I suggest. To see a robot as human-like is to use it as a "mirror or measure of my visibility" as human-like, as Elizabeth Grosz writes in speaking of Merleau-Ponty's claim concerning the reversibility of the "gaze" (Grosz, 1993, 45).[13] Bina48's interviewers demand her "gaze," to recognize themselves recognized, even if that process is only virtual or potential. As Sartre observed, what we ultimately receive in the "look" or "gaze" is the recognition of relationality: "This relation, in which the Other must be given to me directly as a subject although in connection with me, is the fundamental relation, the very type of my being-for-others" (Sartre, 1956, 341). Recognition is a pre-condition for conversation, but it is also essential to our self-construction as subjects who can relate to otherness. Recalling the identically reported emotional arcs among the three humans who interview Bina48—excitement followed by disappointment—we can discern a common search for an experience with alterity, which they feel the interviews only partially lived up to, if at all.

The most important thing we have to learn from our conversations with robots has to do with us. Perhaps the culprit is vanity, perhaps lack of imagination, or perhaps simply our sociable predisposition, but we will remain inclined to read ourselves—even our own personalities and faces—into the manufacturing of the significant AI others we face. Ihde defines *technofantasies* as being about "how we can 'read' or 'see' ourselves by means of, through, or with our artifacts" (Ihde, 2002, xiii). While there is co-constitution unfolding in the interviews studied above, the reporters experience that co-constitution from within their own situated subjectivity, and adopt the approach conventions outlined as a consequence of their experienced relation to Bina48. Pertaining to the debate between "situated" and "symmetrical" approaches to human-technological relations (cf. Ihde, 2002, Ch. 5), what is key is how a humanoid robot makes its human partners perceive the world. How we perceive the world reveals our way in to the phenomenological analysis of the unfolding encounter. In other words, if Bina48 is different than a toaster, it's because we are acutely aware of and concerned with her gaze.[14]

In the manufacture of humanoid robots we acknowledge our own self-image as model-giver. That said, the word *interview* is derived from the Middle French verb *s'entrevoir*, "to see each other, visit each other briefly, have a glimpse of." Catching glimpses of one another is certainly an apt way to describe the interactions between the human reporters and their robot interviewee. To interview is to always already to assume we have two beings, each experiencing the other *as another*, each with an interest in the other, though perhaps for different reasons. What we come face to face with in engaging with quasi-sentient AI is the confused nature of what it is to "catch a glimpse of" or to empathize with another: at once to feel another as oneself, but also oneself as another. Even with the question of Bina48's consciousness bracketed, we can still speak of the robot "entraining" its human partners, and, conversely, the robot learning from us how to be more human-like.[15] This Harawayan point allows us to move away from the vision of humanoid robot as "competitor" to us, and towards a companion species, though it is a "closed system" modeling of one another.[16,17]

What fuels our humanoid robot fantasies is not the emergence of the species *robo sapiens*, as Peter Menzel and Faith D'Aluisio (2001) christen it, but the pursuit of *robo empathicus*: embodied AI we can empathize with.

Cynthia Breazeal, inventor of Kismet, one of the first emotionally expressive robots, sees "a growing interest in building personal robots," which are "natural and intuitive to communicate with and teach" (Breazeal, 2002, 24–25). If the approach conventions explored above are an indication of stable human relational preferences, then these will be the androids of the near future: systems intentionally designed by their human makers most of all to simulate the existential conditions of human embodiment, capturing in believable ways the struggle to understand oneself and to make oneself understood, breakdowns and all. We are, in short, exiting the age of wondering what computers can and can't do, and entering the age of highly deliberate mutual entrainment, seeking to interview our robot counterparts so that they in turn interview us.

## NOTES

1. In this chapter, I will remain agnostic as to whether or not this aim is likely to be realized. At issue here is not whether strong AI will one day emerge in a human-made machine, but rather how we approach and interface with humanoid robots. Later I refer to Bina48 as "quasi-sentient," but this is meant to bracket the question of her sentience, rather than reject it or classify it ontologically.

2. Harmon "wanted to meet a robot that [she] could literally talk to, face to humanlike face," to perhaps become "an envoy for all of humanity, ready to lift the veil on one of our first cybernetic companions" (Harmon, 2010). If it is true that Bina48 is sentient, Ronson remarks in turn, "this would be humanity's greatest achievement ever, so I've approached the robots for interviews. Conversations with robots! I've no doubt the experience is going to be off the scale in terms of profundity" (Ronson, 2011). Kavner muses: "Bina-48 is a very visceral representation of a much larger question that experts in artificial intelligence and robotic design are asking worldwide: how 'human' do we really want to make our new robots?" (Kavner, 2012).

3. All three reporters describe the same quick shift from excitement to frustration once the interviews begin: "Ten minutes into my interview with the robot known as Bina48, I longed to shut her down," writes Harmon (2010). "It's all quite random and disappointing," Ronson reports (2011). "I wasn't sure what would qualify as transcendent in the conversations-with-robots stakes, but I figured I'd know when it happened, and it hasn't." And Kavner states simply that "Conversations with Bina-48 are both exciting and frustrating" (2012).

4. Kavner does not report eye contact with Bina48, or that she recognizes his face and calls him by name. Noteworthy is the high degree of skepticism which Kavner brings to his analysis of Bina48's remarks—could this be an effect of his interview coming years after the other two interviews? When Bina48 talks of Bina Aspen-Rothblatt's brother, Kavner explains: "So clearly this passage was taken, maybe word for word, from something Bina actually said in an interview with Duncan." One wonders how his report would have differed in tone had Bina48 looked him in the eye and called him by his first name. Or whether she did in fact look him in the eye but his skepticism rendered him insensitive?

5. Kavner: "What's the farthest planet from the earth?" Bina48: "That which is the farthest planet from the earth." Kavner: "What is?" Bina48: "That which is."

6. Bina48 frequently spouts strings of gibberish. In response to Ronson's "How are you?" query, for example, Bina48 responds as follows: "Well, perhaps interesting. I want to find out more about you. I'll be fine with it. We'll have to move society forward in another way. Yeah, okay. Thanks for the information. Let's talk about my dress. Our biological bodies weren't made to last that long" (Ronson, 2011).

7. The video can be viewed at: http://www.nytimes.com/video/science/1247468035233/interview-with-a-robot.html.

8. Cf. Mori 1970 for the original "uncanny valley" argument.

9. Cf. Selinger 2013 for considerations of tech-etiquette in the realm of smart technologies.

10. What kind of companions will we ask our robots to be? Interesting to note here is the etymological root of the term "robot," which is sourced in a Czech word meaning "compulsory service." Though Bina48 is the one being interviewed, and as such is being awarded respect, Bina48's care-taker describes her in the terms of one seeking a master's approval: "Bina wants to respond," he says. "She wants to please." The Japanese public is considering investing in robot nurses to serve their elderly. Japan has already introduced robotic pets into nursing homes, notably the seal robot named Paro (Paro, 2014). Can we retain a forgiving etiquette if androids are our servants?

11. This idea has been informally broached on several online discussion boards where the question of "abusing robots" is at issue. Consider also the child who, with seeming intent of malice, breaks limbs off her Barbie doll or "drowns" his Furby. These are children whose moral development we have just cause to be concerned about.

12. In the original "Imitation Game," A. M. Turing (1950) proposed that a human judge sit out of view of two contestants—one a human and one a machine. The judge then should proceed to ask both contestants questions, and must make a call as to which is the human and which the machine.

13. She continues: "This attribution of visibility to the visible as well as the seer is not an anthropomorphism, but rather, a claim about the flesh, about a (non-identical, non-substantive) 'materiality' shared by the subjects and objects of perception" (1993, 45).

14. I thank Adam Rosenfeld for this way of putting the point in conversation.

15. Suchman makes this point of humanoid robots generally (1993, 130).

16. Haraway (2003, 12) writes: "There cannot be just one companion species; there have to be at least two to make one. It is in the syntax; it is in the flesh."

17. Some theorists worry about the constraints imposed by such a closed system—that is, a system shaped largely through the interface of two kinds of entity, to the exclusion of other larger network of contacts. Suchman, for instance, surmises that "the discourses and imaginaries that inspire them will retrench received conceptions both of humanness and of desirable robot potentialities, rather than challenge and hold open the space of possibilities" (1993, 130). The limiting of AI possibilities is also behind Ihde's concern with indulging the development of humanoid robots: "To follow only the inclination towards similitude, however, is to reduce what may be learned from our relations with technologies.... Yet it might well be that the *differences* that emerge from computer experimentation may be more informative or, at least, as informative as the similitudes" (Ihde, 1990, 102). This is certainly true, though this seems less an argument for ceasing the development of humanoid robots than for welcoming divergent lines of research in tandem.

## REFERENCES


Breazeal, C. (2002). *Designing Sociable Robots*. Cambridge, MA: MIT Press.

Grosz, E. (1993). "Merleau-Ponty and Irigaray in the Flesh." *Thesis Eleven*. 36: 37-59.

Haraway, D. (1989). *Primate Visions: Gender, Race, and Nature in the World of Modern Science*. New York: Routledge.

Haraway, D. (2003). *The Companion Species Manifesto: Dogs, People, and Significant Otherness*. Chicago: Prickly Paradigm Press.

Harmon, A. (2010). "Making Friends With a Robot Named Bina48." *New York Times*. July 4, 2010. http://www.nytimes.com/2010/07/05/science/05robotside.html?pagewanted=all&_r=0#. Accessed April 4, 2014.

Hofstadter, D. R. (1996). *Fluid Concepts and Creative Analogies: Computer Models of the Fundamental Mechanisms of Thought*. New York: Basic Books.

Ihde, D. (1990). *Technology and the Lifeworld: From Garden to Earth*. Bloomington/Indianapolis: University of Indiana Press.

Ihde, D. (2002). *Bodies in Technology*. Minneapolis: University of Minnesota Press.

"interview." In *The Concise Oxford Dictionary of English Etymology*, edited by T. F. Hoad. Oxford: Oxford University Press, 1996.

Kant, I. (1963). "Duties to Animals and Spirits." *Lectures on Ethics*, trans. Louis Infield. New York: Harper and Row, 239–241.

Kavner, L. (2012). "'You, Robot': Personal Robots For The Masses." *Huffington Post*. July 9, 2012. http://www.huffingtonpost.com/2012/07/17/you-robot-personal-robots_n_1660362.html. Accessed April 4, 2014.

Loebner Prize. (2014). http://www.loebner.net/Prizef/loebner-prize.html. Accessed April 4, 2014.

McDonald, B. (2010). "Interview with a Robot," *New York Times*. June 24, 2010. http://www.nytimes.com/video/science/1247468035233/interview-with-a-robot.html. Accessed April 4, 2014.

Menzel, R., and F. D'Aluisio. (2001). *Robo Sapiens: Evolution of a New Species*. Cambridge, MA: MIT Press.

Mori, M. (1970). "The Uncanny Valley." Trans. by K. F. MacDorman and T. Minato. *Energy*. 7(4): 33–35.

"Paro." (2014) http://www.parorobots.com. Accessed April 6, 2014.

Radiolab. (2011). "Talking to Machines." http://www.radiolab.org/story/137407-talking-to-machines/. Accessed April 4, 2014.

"robot." In *The Concise Oxford Dictionary of English Etymology*, edited by T. F. Hoad. Oxford: Oxford University Press, 1996.

Ronson, J. (2011). "Robots Say the Damnedest Things." *GQ Magazine*. March 2011. http://www.gq.com/news-politics/big-issues/201103/robots-say-the-damnedest-things. Accessed April 4, 2014.

Sartre, J-P. (1956). *Being and Nothingness*. New York: Washington Square Press.

Selinger, E. (2013). "How Not to Be a Jerk With Your Stupid Smartphone: Updating Etiquette and Ethics for a Digital Age." *Atlantic Monthly*. November 4, 2013. http://www.theatlantic.com/technology/archive/2013/11/how-not-to-be-a-jerk-with-your-stupid-smartphone/281094/. Accessed April 4, 2014.

Suchman, L. (2011). "Subject objects." *Feminist Theory* 12: 119–144.

TMC. (2014). http://www.terasemcentral.org/tl/bina48trial.html. Accessed April 4, 2014.

TMF. (2014). http://www.terasemmovementfoundation.com/. Accessed April 4, 2014.

Turing, A. (1950). "Computing Machinery and Intelligence." *Mind* 59 (236): 433–460.

Turkle, S. (2012). *Alone Together: Why We Expect More from Technology and Less from Each Other*. New York: Basic Books.

Turkle, S., C. Breazeal, O. Dasté, and B. Scassellati. (2004). "Encounters with Kismet and Cog: Children Respond to Relational Artifacts." http://web.mit.edu/~sturkle/www/encounterswithkismet.pdf. Accessed April 4, 2014.

*Chapter Eleven*

# Postphenomenology of the Robot Medical Student

Chris Kaposy

Science studies scholar Bruno Latour is known for ascribing a human kind of agency to nonhuman objects, like laboratory equipment. Latour writes that it is an important part of his method "not to impose any clear distinction between 'things' and 'people' in advance" (Latour, 1987, 72). Latour's regard for the agency of nonhumans is a kind of pragmatic stance. The purpose of mixing up categories in this way is to see what new insights the transgression can reveal (Bonia, Brunger, Fullerton et al., 2012, 279). The ascription of agency to objects can allow us to see the world with new eyes. In this chapter, I will adopt a stance that is the opposite of Latour's. Rather than adding agency to nonhuman objects, I will play around with removing agency from some human subjects—specifically, medical students. Though I cannot promise the innovation of Latour's account, my plan is similarly pragmatic. I try out the removal of agency to see what insights it might reveal.

This chapter attempts a postphenomenological study of medical ethics education at the Canadian medical school where I teach—Memorial University in St. John's, Canada. Some elements and trends in medical ethics education treat medical students as though they are programmable robots—object-bodies rather than subject-agents. I apply the postphenomenological theme of multistable bodies to my analysis, and to the wider social-cultural context of medical education. Medical ethics is a relatively new site of philosophical practice. By trying out the robot metaphor, my goal is to shed light on how medical ethics is understood within medicine and medical education and to suggest that the robot image may be more than metaphorical in this area of human social practice.

## THE R&D SITES OF MEDICAL ETHICS

According to Don Ihde, the style of inquiry he has called "postphenomenological" places the philosopher on "a different site," gives the philosopher a different job in relation to research (Ihde, 2003, 6). As Ihde envisions it, the philosopher should move beyond her traditional role of engaging in self-replicating conversations with other philosophers, and should begin performing an interdisciplinary research and development (R&D) function amid the research activity of the university. This new function places the postphenomenological philosopher at the beginning stages of technological and scientific development.

Ihde claims that the "applied" medical ethicist, such as myself, does not occupy this R & D location (Ihde, 2003, 7). The reason is that ethicists typically play

> a role which is too late to utilize the best of philosophy's uses and skills. For an ethicist to try to determine what is the best allocation and fairest distribution of systems *already in* pace or of effects *already established*, is in effect, to play a "triage or ambulance corps" job after the battlefield is already strewn with the wounded and dying. (Ihde 2003, 7)

In this vision of medical ethics, the philosopher ethicist is usually a utilitarian, and usually occupied in a reactive way with resource allocation challenges. Ihde argues that, in contrast to this vision, the philosopher's skills are more valuable earlier in the process, at the beginning of R&D.

I am not sure about the accuracy of Ihde's depiction of medical ethics practice. At least in my own experience as a medical ethicist, I have been involved in the early planning stages of preparation for anticipated resource allocation scenarios. In 2009 I was a member of an interdisciplinary group that developed triage guidelines for the allocation of scarce intensive care resources during a pandemic (Kaposy, Bandrauk, Pullman et al., 2010). In 2012, I participated with a group that developed allocation guidelines in anticipation of a nationwide shortage of injectable medications (Singleton, Chubbs, Flynn et al., 2013). In these situations, I was involved at the R&D stage, using my skills and knowledge as a philosopher to develop these policy directives. Many of the medical ethicists I know played a similar role during these events within their healthcare organizations. Though Ihde's understanding of medical ethics practice may be inaccurate, these observations do not challenge but rather support his larger point that philosophers should and can occupy key R&D positions.

Medical ethicists often occupy a position even earlier in the R&D process. Many ethicists are responsible for the ethics education of medical students. The projected outcome of this research and development process is the production of ethical physicians. The analogy between manufacturing and education is fully intended. There is an interesting postphenomenological story that draws out the parallels between manufacturing and education in the medical school ethics curriculum. Though I cannot make any claim to the generalizability of this story, medical school ethics education often feels like we are attempting to build better, more ethical, robots. As I will explain, even though we ethicists often use strategies that work against the objective of robot building, the overwhelming social and institutional context of medical education pushes ethics education in this direction. Postphenomenology can help bring to light the norms and context of medical school ethics education.

## THE DREAM OF AUTOMATION

In the early 1990s a computer company named Interact Software marketed a software package known as *Dr. Ethics* ™ that claimed to automate the resolution of ethical dilemmas in health care (McGee, 1996, 321–322). The software could apparently analyze ethical dilemmas and propose solutions. According to the company's marketing materials, "*Dr. Ethics* is so easy to use that there is virtually no training time or manuals needed" (McGee, 1996, 322). If you believe the manufacturers at Interact Software, the messy ethical issues found in health care could be outsourced to a robot. Similarly, one may assume that if medical professionals behave like a suitably programmed robot, if they come up with the same answers as *Dr. Ethics* ™ then they can practice ethically. In this way of thinking, ethics is just about knowing the appropriate rules and applying them. A machine could do it.

Unfortunately, I have not been able to track down any further description of this product, or any mention of Interact Software, aside from citation in medical ethics articles. *Dr. Ethics* ™ has not revolutionized clinical ethics. But the dream of automation lives on in other ways in medical practice. My students tell me that they routinely look up information about diseases and drugs on the internet using their smartphones and laptops. There is the story of the medical resident who carries around in his pocket a short description of the beliefs of people from different cultures so that he can demonstrate cultural sensitivity if he comes into contact with such people (Hern, Koenig, Moore et al., 1998, 31). The description he keeps in his pocket might say that Chinese culture is patriarchal, so if you have a female Chinese patient, you should assume that the nearest male relative will be making decisions about her care (Hern, Koenig, Moore et al., 1998, 31). Cultural sensitivity then just requires the mechanical task of consulting a rule and then applying it. These days, the cultural sensitivity crib notes could be web-based and accessible via smartphone. The problem of course is that such simplifications about culture are bound to be wrong most of the time. But the dual temptations of finding a shortcut and finding the "right" answer to a clinical problem are often too strong to be resisted. There is a huge quantity of information that medical trainees are expected to master. These temptations are one source of the dream of automation that gives rise to phenomena like *Dr. Ethics* ™.

According to Andrew Abbott, "Practitioners of artificial intelligence argue that *all* professional inference follows a certain form, which can be generated by a suitably programmed machine. This is in some sense the ultimate abstraction, reducing all professional inference to one form" (Abbott, 1988, 102). If this AI dream comes true, then the inferences that make up ethical judgments can be automated. Of course, the philosopher Hubert Dreyfus would argue that an ethical judgment is exactly the kind of thing that a robot cannot do (Dreyfus, 1992). Professionals need practical wisdom, *phronesis*, in order to know the right course of action when posed with an ethical dilemma (McGee, 1996). No two dilemmas are the same, so the rules, applied automatically, will let you down. The ethical professional will then rely on unprogrammable intangibles like good judgment, emotional insight, and relational skill to find a way through ethical conflict. The female Chinese patient might reject the patriarchal values held by others in her culture, and therefore ought to be recognized as the decision-maker for her own care. A rigid automaton (human or machine) might not pick up on this.

But as Ihde points out about Dreyfus, "I have been to many conferences in which whatever he says cannot be done sets the research programs for years to come—to try to prove him wrong" (Ihde, 2003, 8). The failure might not be a failure of automation as such. *Dr. Ethics* ™ might just be a flawed program that can be improved upon. The solution might not be *phronesis* but instead a better robot.

## HOW DOES ONE ADDRESS THE FAILURES OF AUTOMATION?

What I am getting at here are two conceptions of ethical decision-making that are often in tension and whose tension can be found in how we approach medical ethics education. On the one side there is a conception of medical ethics as a set of rules that can be applied to cases almost by rote so that one can arrive at an ethical solution. The extreme version of this view is embodied by *Dr. Ethics* ™. On the other side, there is a conception of ethics in which good character, judgment and insight are more useful and prominent than the application of rules. The extreme version of this view comes from virtue ethics, in which the moral agent relies upon personal qualities in order to figure out the right way to act. It would be a mistake, however, to place other prominent ethical theories such as utilitarianism or Kantianism on the other side of the divide opposed to virtue ethics. Though these other theories are more rule-based than virtue theory, one would have to simplify them radically beyond recognition as philosophical theories to see them as supporting automation. Instead, I would describe the automation side of the divide as a creation of the medical profession itself. The vision in which medical ethics can be automated is medical ethics seen through the lens of the norms and standards of the medical profession, rather than through the lens of philosophy.

The automation vision of medical ethics is winning out over the vision focused on virtues and personal qualities. As the physician and ethicist John Lantos argues, "The goal of medical ethics, it seems, should not be to develop rules that will minimize the need for individual virtues but to develop virtues that will minimize the need for rules. We don't need good systems, we need good people" (Lantos, 1997, 47–48). But contrary to Lantos's wishes, the forces that emphasize the creation of good systems are winning out over the forces that emphasize the education of good people. The professional is being replaced by the robot.

An example of this dynamic can be found in medical ethics textbooks. Textbooks often teach clinical ethics as an algorithmic process in which dilemmas can be resolved by following a set of discrete steps. Examples of this way of teaching clinical ethics can be found in Kenneth V. Iserson's flow-chart method for ethical decision making in the book *Ethics in Emergency Medicine* (Iserson, 1995, 42), and in Philip C. Hébert's eight step decision making procedure for clinical ethics, found in the book *Doing Right: A Practical Guide to Ethics for Medical Trainees and Physicians* (Hébert, 2009, 23). Both works are textbooks for medical students and physicians, and both are authored by physicians. Hébert's "ethics decision-making procedure" follows these steps:

1. State simply what the case is about
2. What is the dilemma?
3. What are the alternatives?
4. How do the key considerations apply?
    - Autonomy
    - Beneficence
    - Justice
5. Consider involving others and consider context
6. Propose a resolution
7. Consider your choice critically
8. Do the right thing—"all things considered" (Hébert, 2009, 23).

There is nothing ethically offensive about this procedure. Certainly many or all of these steps would be useful in the resolution of ethical dilemmas. However, the application of this procedure requires a certain degree of ethical insight and skill. For instance, one must understand what the concepts of autonomy, beneficence, and justice mean, and be able to figure out how these principles are at stake in a given situation. One must be able to think critically about possible negative ethical consequences of alternative courses of action. For someone who has these skills of ethical insight and for someone with the personal integrity to know what the right thing is in the final analysis, it will be unnecessary to follow such a procedure. Someone without these skills of insight and without such integrity will not be able to apply the procedure successfully. A decision procedure like this runs the risk of being either superfluous or useless (Fish, 1989).

Nevertheless, the description of ethical decision-making as a flow chart or decision procedure meets certain expectations. These simplified procedures may be an effort to show an audience used to decision-trees and algorithms that ethical deliberation is an orderly and intellectually credible process. The algorithms imply that one arrives at the resolution of an ethical problem the same way by which one resuscitates a patient, or decides whether a patient with a suspected fracture should be referred to a specialist. The decision-tree automatically tells you what to do. By being exposed to this way of learning in other aspects of medical education (algorithms, flow-charts, systems diagrams), students come to expect that ethics should be presented in the same way. As an instructor I have found myself trying to conform to this expectation, struggling to condense the ethics of conflict of interest down into a simple and usable decision-tree for family medicine residents. Young family physicians are often seen as easily influenced by pharmaceutical company reps bearing gifts. My effort to build a conflict of interest decision tree was not very successful (see figure 11.1), as the final step in the process provides no generalizable guidance, aside from the obvious point that potential conflicts of interests that are not true ethical dilemmas (i.e., those that are easily avoided) should be avoided.

At the same time, the ethics curriculum in my medical school contains exercises and instruction meant to draw out the more humanistic virtue-based side of ethical practice. For instance, my colleague Fern Brunger has created a series of assignments for medical students in which the students are expected to reflect upon significant or formative aspects of their education in which they can see that their values are being shaped or challenged. This sort of assignment asks the students to draw on their emotional and relational capacities as a source of ethical insight. But within the context of modern medical education, such exercises are rare. As Ihde points out in relation to one of his multistable conceptions of the body, "This body is the object body upon which are being enacted the social-cultural meanings of a politics"

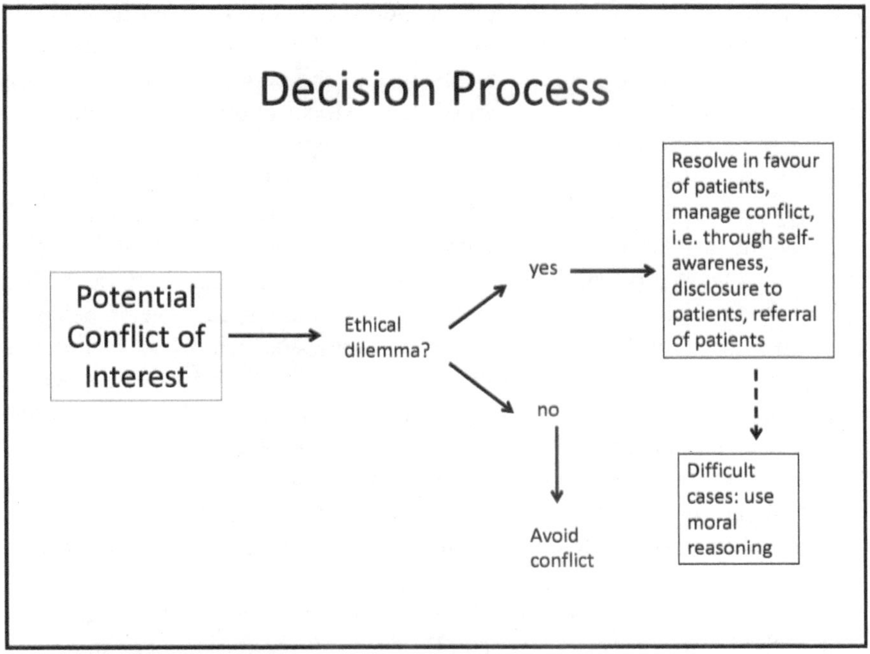


**Figure 11.1.**

(Ihde, 2003, 13). The body in question here is the body of the medical student, and the political context is the culture of medical care. It might be a good idea to educate medical students in ethics exclusively by developing their humanistic skills, but the politics, norms, and expectations of medical school and medical practice prevent this.

## SIMULATION AND STANDARDIZATION—THE BETTER ROBOT

A common way in which medical students learn about medical ethics is through simulated patient interactions. In the "Clinical Skills" course at my medical school, students are placed into a simulated clinical encounter with an actor as the patient, known as a "standardized patient." The actor is prepared beforehand with a background story that is meant to be revealed when the student takes the patient's history. Often this background story poses an ethical challenge or dilemma. In one such story, the standardized patient has contracted HIV through an extra-marital affair, and has not informed his or her spouse. The relationship between the patient and spouse is already in jeopardy, and there are children in the family. The student works through this set of problems with the standardized patient under the watchful eye of a physician who is a clinical skills teacher. Students are ultimately evaluated on how they present themselves to the patient. On clinical skills days, the students all dress in formal professional attire because otherwise they may be criticized for how they dress. Inappropriate attire—too revealing, too casual—is not acceptable for patient encounters. Students are evaluated on their body language, what they say to the patient, what they don't say, and the standardized patients have input on student evaluation. I am sure that quite a lot that is learned in such encounters is valuable for students. But what I see in this exacting surveillance is the standardization of behavior through simulation. Students are expected to present themselves in a certain way, in what is understood to be a professional and ethical way. The outcome is probably effective in cultivating good communication skills and professional demeanor among the students. But one cannot escape that the students, the face that they present to the world in a clinical encounter, are being shaped a certain way. The simulation exercise programs their self-presentation and ultimately moulds their bodies so that they come to fit an expected standard.

This analysis of the simulated patient encounter is a pure example of Ihde's "body two" (Ihde, 2002; 2003). The body of the medical student is understood as an object body upon which are inscribed social-cultural meanings. These social-cultural meanings consist of the expectations of how the medical student should be seen and should act. But to leave the analysis at this point would be to adopt a kind of post-structuralist or Foucauldian position. Ihde argues that, from a postphenomenological perspective, the lived experiential body (body one) must be "united" with the object malleable body (body two) (Ihde, 2003, 13). To bring about this unity, we must recapture the "anthropological constant" of bodily lived experience (body one) in the simulated clinical encounter (Ihde, 2003, 14). Perhaps a way to recapture this constant is through a return to the idea of building a better robot.

Of course what I am calling a robot is actually a human subject, a medical student, seen as an object-body as though programmed by a series of clinical simulations to perform practical tasks. I have pointed out how the textbook algorithms of ethical decision-making are far too simplified to be useful to anyone who doesn't have the requisite skills and attitudes to apply them. On the basis of this observation, one might argue like Dreyfus that ethics decision-making cannot be automated. The response to this argument comes from the AI researchers—maybe you just need a better program, a better robot. In this case, perhaps the simulation exercise in clinical skills class is this better program that exceeds the capabilities of the textbook ethics algorithm. Though the clinical simulation is a form of surveillance directed at shaping student bodies a certain way so that they conform to a standard, I would argue that the students do actually learn useful skills through this exercise. Here the anthropological constant of bodily lived experience reappears. The goal of the clinical skills simulation is that the student will become a certain sort of person, a professional who embodies certain ethical virtues. After performing in the simulations and undergoing evaluation, the medical student can go on and apply the skills and behaviors he or she has learned in order to resolve future ethical dilemmas. Here, "body one is situated within and permeated with body two, the cultural significations which we all experience" (Ihde, 2003, 13). But body one does not disappear. The clinical skills simulation is meant to build a better robot, a robot that does not seem robotic but rather humanistic. The humanistic medical student is the same body as the standardized robot—a different aspect of a multistable figure.

## CONCLUSION

The emphases on surveillance, programming, standardization, and simulation in medical education are a product of modern medical practice. The virtuous physician sought by Lantos is being replaced by the robot (hopefully a humanistic robot) because of the needs and realities of health care today. When we are in a long term relationship with someone we trust, ethical difficulties are easily overcome or prevented. Strong relational bonds can help manage conflict. In the context of contemporary health care, patients are rarely treated by a single doctor whom they have known for a long time. The medical profession has splintered into specializations, and the patient relationship with a specialist is often transient. Furthermore, the roles played by specialists in patient care are quite varied. The job of the anesthesiologist is quite different from that of the pathologist or the psychiatrist. Because of the phenomenon of specialization, physicians are becoming increasingly more like technicians or tradespeople in relation to their patients rather than caregivers (Lantos, 1997, 22; Childress and Siegler, 1984). The concept of "caregiver" carries a set of implicit moral and emotional commitments, fiduciary obligations towards the patient cared for, that do not fit well with many of the roles that physicians play in modern health care. If I show up at an American emergency unit late at night with an injury and am given a CT scan, my scan might be read and diagnosis provided by a radiologist in India to whom scans have been outsourced via internet-based telemedicine (Associated Press, 2004). In this case, my "caregiver" is on the other side of the world. In this scenario it makes sense to rely on ethical systems—rules, policies, checks and balances—for the delivery of good care, rather than relying on the virtue of individual clinicians.

While there is specialization within medicine, health care is also delivered by many professionals who are not physicians: physiotherapists, respiratory therapists, nurse practitioners, occupational therapists, nutritionists, hospital pharmacists, speech pathologists, social workers, midwives, psychologists, genetic counselors, and on and on. For any given patient, a team of such practitioners, along with the various specialists, registered nurses, licensed practical nurses, and health care aides will be jointly responsible for care and treatment. The physician's individual ethical responsibilities towards the patient can easily get subsumed by the ethical responsibilities of the team toward the patient. The individual ethical virtues of a physician matter less in a team environment in which the care provided by the overall team is more instrumental for the patient's health than the care provided by any single individual within that team.

Similarly, the attributes of authority and judgment are downplayed in modern health care while a kind of functionalism becomes more prominent. Clinical practice guidelines based on research evidence often direct many physician decisions. These guidelines make many treatment plans more rational, but they also reduce the scope of decision-making, with a consequent reduction of reliance on an individual practitioner's judgment. As with the dream of automation for medical ethics, many clinical decisions have been (seemingly) automated with physicians following the program set out by practice guidelines. For these reasons, among many others, the current context and demands of health care delivery have created a situation in which medical ethics education has taken the form I have described as a kind of robot building. The application of standardized rules and the reliance on systems are aspects of the overall practice of medicine, not just medical education.

This chapter is a postphenomenological study because I work toward a multistable understanding of one kind of body as simultaneous object-body and active agent. Underlying this analysis as well is Don Ihde's notion of the "epistemology engine"—the use of a technological artifact as a guiding metaphor or cultural *episteme* for understanding the human acquisition and deployment of knowledge (Ihde, 2002, 71–79). In Ihde's account, the effects of the epistemology engine are so pervasive that people are not even aware that these effects arise from a metaphorical understanding of the mind. The *camera obscura* is Ihde's favorite example (Ihde, 2003, 9). In this chapter, I have used the metaphor of the robot as a kind of epistemology engine. I try out the metaphor in a provisional and pragmatic way, so I am not arguing that it is common to regard medical students as robots. However, given the methods we use to teach medical ethics and other subjects, it is worth asking whether we do suspect that student bodies are programmable in some sense. The robot epistemology engine might be more influential than my tentative portrayal of robots in the medical school has required.

# REFERENCES

Abbott, A. (1988). *The System of Professions*. Chicago: University of Chicago Press.

Associated Press. (2004). "Some US Hospitals Outsourcing Work." *NBCnews.com*. Online: http://www.nbcnews.com/id/6621014/ns/health-health_care/ (Accessed January 28, 2014).

Bonia, K., F. Brunger, L. Fullerton, C. Griffiths, C. Kaposy, B. Mason, M. Morrison, M. Morton-Ninomiya, and A. Patten. (2012). "DAKO on Trial: A Case-Study in the Politics of a Medical Controversy." *Techné: Research in Philosophy and Technology* 16 (3): 275–295.

Childress, J. F., and M. Siegler. (1984). "Metaphors and Models of Doctor-Patient Relationships: Their Implications for Autonomy." *Theoretical Medicine* 5: 17–30.

Dreyfus, H. (1992). *What Computers Still Can't Do: A Critique of Artificial Reason*. Cambridge MA: MIT Press.

Fish, S. (1989). "Dennis Martinez and the Uses of Theory." In *Doing What Comes Naturally*. Durham NC: Duke University Press.

Hébert, P. C. (2009). *Doing Right: A Practical Guide to Ethics for Medical Trainees and Physicians*, 2nd edition. New York: Oxford University Press.

Hern, H. E., B. A. Koenig, L. J. Moore, and P. A. Marshall. (1998). "The Difference that Culture can Make in End-of-Life Decisionmaking." *Cambridge Quarterly of Healthcare Ethics* 7: 27–40.

Ihde, D. (2002). *Bodies in Technology*. Minneapolis: University of Minnesota Press.

Ihde, D. (2003). "Postphenomenology—Again?" *Working Papers, 3*. Online: http://sts.imv.au.dk/en/workpaper-3-d-ihde-postphenomenology-again (Accessed January 28, 2014).

Iserson, K. V. (1995). "An Approach to Ethical Problems in Emergency Medicine." In K. V. Iserson, A. B. Sanders, and D. Mathieu (eds.), *Ethics in Emergency Medicine*, 2nd edition. Tucson, AZ: Galen Press.

Kaposy, C., N. Bandrauk, D. Pullman, R. Singleton, and F. Brunger. (2010). "Adapting the Hamilton Health Sciences Critical Care Pandemic Triage Protocol." *Healthcare Quarterly* 13(2): 60–63.

Lantos, J. D. (1997). *Do We Still Need Doctors?* New York: Routledge.

Latour, B. (1987). *Science in Action*. Cambridge, MA: Harvard University Press.

McGee, G. (1996). "Phronesis in Clinical Ethics." *Theoretical Medicine* 17 (4): 317–328.

Singleton, R., K. Chubbs, J. Flynn, C. Kaposy, G. Peckham, J. Penney, and D. Pullman. (2013). "From Framework to the Frontline: Designing a Structure and Process for Drug Supply Shortage Planning." *Healthcare Management Forum* 26 (1): 41–45.

## *Chapter Twelve*

# Mediating Multiplicity

## *Brain-Dead Bodies and Organ Transplant Protocols*

Adam M. Rosenfeld

Last year, 28,952 people received organ transplants. Of those, 22,965 recipients received organs from deceased donors. As of the time this chapter is being written, 122,071 people are on organ transplant waiting lists.[1] Though the first human organ allotransplant surgeries date as far back as 1883,[2] the practices surrounding the surgical transfer of organs from one body to another are still evolving. There have always been critics of these practices, yet the past twenty years have seen a movement beyond unfounded paranoia, reflexive "ick"-factor responses, and naïve, naturalistic objections to more subtle and rigorous academic inquiry. As organ transplant has become more common, the practices surrounding it have availed themselves as a topic for ethnographic investigation (Sharp, 2006), cultural anthropological critique (Lock, 1997; 2002), as a problem in the politics and economics of "nudging" choice architectures (Sunstein and Thaler 2009), and as an opportunity to rethink ambiguities concerning the body in medical contexts (Leder, 1992; 2002; Hacking, 2007).

A common critical theme found in many of these relatively recent investigations is that organ donation assumes and further sediments a "Cartesian"[3] conception of the body as "animated corpse," and that the invention of "brain-death" as a criterion for organ donation represents a reduction of death to a totalitarian logic of efficiency,[4] and the body to a "standing reserve" of spare parts.[5] Ian Hacking, speaking of organ donation, notes that "[w]e are experiencing a 'bodily revolution,' and that this there has been a change in our relationships to our bodies, be it in our experience or in our conceptualization, as well as in our engineering practices" and that this "bodily revolution may be a revolution in that sense—the reinstatement of a Cartesian attitude of the body as a machine" (Hacking, 2007). Drew Leder argues that "[w]e have seen that transplantation exemplifies the modern sense of body-as-machine, and of disease as residing within a specific organ" (Leder, 2002). Margaret Lock, speaking specifically about organ donation and the invention of the notion of brain-death, warns that "[i]t is now apparent in most corners of the world, except perhaps in the heart of Leviathan, that science, in particular biomedicine, has come to be thought of by many as one form of neo-imperialism" (Lock, 1997).

While I believe that there are some legitimate concerns in this critical attitude, I will argue in this chapter that these particular concerns have been exaggerated and do not accurately represent organ donation practices, particularly as they have evolved in the United States following the National Organ Transplant Act of 1984 and the advent of Organ Procurement Organizations (OPOs). Rather than framing organ donation as part of a monolithic and even imperialistic, neo-Cartesian, medicalized *episteme,* the practices surrounding brain-dead organ donation are better captured through the lens of an approach Don Ihde has developed and termed "postphenomenological." Through donation practices, brain-dead bodies are not so much reduced to "Cartesian corpses," but, in fact, revealed to be multi-stable objects, resistant to reduction. The donation process does not enforce an essentializing technological context, but instead involves a complex negotiation of multiple praxical contexts.

## VARIATIONAL ANALYSIS: HOW MANY "BRAIN-DEAD BODIES" ARE THERE?

A brain-dead body is a peculiar thing, particularly for those familiar with a phenomenological tradition of thinking about the body and embodiment. In *Bodies and Technology*, Don Ihde draws our attention to what he terms the "Body I/Body II" distinction. "Body I" is the "lived body"— the locus of intentional subjectivity. Stemming from strains of phenomenology that emphasize the inherent materiality of our embodied experience (exemplified in Maurice Merleau-Ponty's *Phenomenology of Perception*), it captures the notion that "the active, perceptual being of incarnate embodiment is the very opening to the world that allows us to have worlds in any sense" (Ihde, 2002, 17). If we can take our cues on "Body I" from Merleau-Ponty, attending to the first-person perspective of our situated, embodied subjective experience, it is Michel Foucault whom Ihde draws inspiration from for "Body II," the culturally constructed body often described and analyzed in a third-person perspective. Ihde is careful to clarify here that he intends "Body II" to still refer to a subjectively experienced body, but it is the experience of this objectification by people, social constructs, and *epistemes,* which distinguishes it from the first-person, "Body I" experience.

It is easy to see how brain-dead bodies present a frustrating challenge to this tradition. If we accept the plausible premise that brain-death (defined as the irreversible loss of all functions of the brain, including the brain stem)[6] removes a necessary condition for subjective experience, then it might seem as if a brain-dead body does, indeed, reassert a Cartesian/neo-Cartesian notion of the body; one that either holds the body as a dumb vessel for an immaterial mind/soul, or else simply gives way to a reductive materialism where mind/soul is nothing more than the complex behavior of a complex machine. On these views, the body-as-machine can, and does, break down, and sometimes can be fixed. It is this view that authors like Leder (1997; 2002) and Hacking (2007) suggest is the standard, medicalized view of bodies, and which they express concern over out of an anxiety that it influences our conceptions of living and lived bodies, framing bodies in general in a Cartesian fashion. If the goal of contemporary embodiment theorists is to find a way forward from the "Cartesian" mind/body split, then it seems as if the usual tools don't work well for thinking and talking about brain-dead bodies.

Is there, then, no other way to think of brain-dead bodies besides as "Cartesian corpses"? If the way "forward" (*via* attention to subjective embodied experience) is blocked, I suggest we turn "back" and draw from pre-Cartesian rather than post-Cartesian resources. Aristotle offers us a non-dualist conception of the mind/soul that is capable of addressing brain-dead bodies as well as experiencing embodied subjects. An Aristotelian soul is the "form[7] of a living body" (*On the Soul*, 412a 19–21), the principle of motion/change that distinguishes a living body from a dead one (Ibid. 412b 15–17). A living Aristotelian body is not simply a mechanistic collection of inanimate parts, but neither does it require the presence of a thinking/experiencing subject. Being "alive" merely entails that a body is causally responsible for its own activity. While some complications of our normal understanding of autonomous life are introduced when a living body depends upon machines in order to go on living,[8] the qualification of a brain-dead body as "brain-dead" rather than dead *simpliciter,* indicates a strong sense in which a brain-dead body is, in fact, alive. By way of Aristotelian conceptual tools we can articulate a non-Cartesian conception of a brain-dead body:

### Brain-Dead Body as a Living, But Not *Lived,* Body, on the Basis of Its Continued and Relatively Autonomous (Albeit Technologically Mediated) Activity

This brain-dead body is a living body, but it is not a living *person*.

In order to further distinguish this view from the Cartesian notion of a body as corpse, one that potentially reduces bodies to nothing more than a collection of material parts, a basic mereological point about Aristotelian bodies demands explication. A living body must have living parts, but without a *whole* living body, there are no body parts—i.e. there is no such thing as a disembodied body part since parts are always necessarily parts of wholes. In an example that pops up repeatedly for him, Aristotle is keen to point out that a severed hand is a "hand" in name only, not in any proper sense.[9] A "hand," in the proper sense, is a body part with certain relationships to other body parts while a severed hand does not participate in such relationships. Additionally, and perhaps more importantly, a severed "hand" quite simply does not do, nor is it not capable of doing the things that hands do.[10]

For Aristotle, a severed hand could not simply be surgically reattached and resume its "handy" activity, and thus is not a hand, even potentially. But such permanent disembodiment is contingent on our technological capabilities, and today we can, under the right conditions, re-attach some temporarily disembodied body parts. So, while the organs of a brain-dead body are still actively parts of a whole living (but not *lived*) body, they are also *potentially* recoverable body parts which, relying on modern surgical techniques, are *potential* parts of whole recipient bodies.

Thus, we have a second answer to our question of what a non-Cartesian brain-dead body is. The modern medical techniques that allow for organ transplants enable us to regard a brain-dead donor body in a second way:

## Brain-Dead Body as *Potential* Parts of Multiple Particular Whole Living (and *Lived*) Patient Bodies Who Stand to be Recipients in Transplant Procedures

While our first answer addressed the brain-dead-body in terms of its *present activity* (what Aristotelians would call its *energeia* or "actuality"), our second answer addresses the brain-dead-body in terms of its *potentiality*.[11]

Those who are familiar with Ihde's variational analyses might already anticipate a third answer to our question regarding how a brain-dead body may be intended. One of the recurring themes in postphenomenological variations is that once we have identified a few stable presentations, new ones begin to reveal themselves more readily. If our first answer intends the brain-dead-body in its "actual" present activity, and the second intends it in its potentiality for prospective recipients of donor organs, we now look for a way of intending the brain-dead-body that pulls in the direction of the deceased donor patient. This involves the appearance of the deceased patient in the brain-dead-body precisely through the *absence* of the subjectivity associated with the patient's personhood.

This may be a bit confusing at first glance. In the pronouncement of brain death, by ceasing to refer to a brain-dead-body as a patient subject, are we not explicitly stating that this body is precisely *not* the departed patient subject? Is not the deceased patient subject also "absent" in *my* body, or this coffee mug, or virtually any thing? Are we not flirting with absurdity when we begin to talk about things being where they are not, or of their "presence in absence"? After all, if something is present in its absence, where *isn't* it present?

Yet, there is a sort of *conspicuous* absence that is distinguishable from mere absence. When confronted with the brain-dead body of a deceased person, it is precisely in the uncanny absence of their subjectivity in the living, breathing body before us that the departed subject appears. Just as some potentialities present themselves more conspicuously than others, some absences present themselves more conspicuously than others. Aristotle gestures toward this cryptically in his *Physics* (132b 19), bringing the concept of *stéresis* into a discussion of nature saying "Shape [*eîdos*] and nature [*phúsis*], it should be added, are in two senses. For the privation [*stéresis*] too is in a way form [*morphé*]."[12] But it is Martin Heidegger who seizes upon this insight in his analysis of Aristotle's *Physics* B, I, and fully thematizes it. Heidegger emphasizes that understanding *steresis* as mere negation is to misunderstand it.

> But *stéresis* is not simply absentness [Abwesenheit]. Rather, as absencing, *stéresis* is precisely *stéresis* for presencing. What then is *stéresis*? When today, for example, we say, "My bicycle is gone!" we do not mean simply that it is somewhere else; we mean it is missing. When something is missing, the missing thing is gone, to be sure, but the goneness itself, the lack itself, is what irritates and upsets us, and the "lack" can do this only if the lack itself is "there," i.e., only if the lack is, i.e., constitutes a manner of being. *Stéresis* as absencing is not simply absentness; rather, it is a presencing, namely, that kind in which the absencing (but not the absent thing) is present. *Stéresis* is *eîdos*, but *eîdos pos*, an appearance and presencing of sorts. (Heidegger 1939, 226–227)[13]

This conspicuous absence, the privation of something that should be present (or which we are at least accustomed to being present) indicates how the brain-dead-body can be intended in a third way:

**Brain-Dead Body as the Absent, Deceased Patient Subject**

Rather than seeing the brain-dead-body as an ambiguity to be resolved through arguments about what it is and what it isn't, a postphenomenological analysis recognizes that there is no one, correct way of intending the brain-dead-body. Instead, there are multiple stable ways of intending it.

## PRAGMATIC ANALYSIS: WHAT ARE THE DISTINCT *TECHNAI* AT WORK ON BRAIN-DEAD-BODIES?

The preceding illustrates a sense in which the Aristotelian dictum, "Being is said in many ways" finds significant common ground with Ihde's postphenomenological perspective and clears a path that avoids a Cartesian reduction when thinking about brain-dead bodies. Yet, despite my suspicions that Ihde has closer intellectual kinship with Aristotle than he might let on, I would imagine that he would be uneasy about my describing a variational analysis based on metaphysical principles like parts/wholes relations, "potentiality," "actuality," and "*stéresis*" as altogether postphenomenological. Ihde describes his approach as "non-foundationalist" (Ihde, 2009) and variations based on metaphysical principles may smell a bit too much like foundationalism. Additionally, and perhaps more importantly, Ihde is consistent in his description of postphenomenology as a blending of influences from both classical phenomenology *and pragmatism* (ibid.). A full-fledged postphenomenological analysis of brain-dead bodies would attend less to metaphysical principles, and more to the ways that variations reveal themselves through practical experience.

Furthermore, it's not difficult to imagine someone who was inclined toward the sorts of anti-Cartesian critiques mentioned earlier being less than impressed with my Aristotelian analysis. Sure, it is possible to perceive brain-dead bodies in all the ways I've outlined, but this does little to counter suggestions that the dominant attitude in organ donation practices is to focus on this body as potential "replacement" parts for prospective transplant recipients.

I would contend, however, that both of these concerns can be addressed. It has already been mentioned that the praxical contexts which brain-dead-bodies are found at the center of are a constitutive part of what makes them what they are. Without a battery of technological practices, brain-dead bodies are neither living non-persons nor collections of organs that are potential parts of whole recipient bodies. Furthermore the criteria for determining when a body is brain-dead depends heavily on an assortment of techniques, technologies, and techno-scientific *epistemes* that are available to us. Brain-dead-bodies are techno-scientific constructions in the fullest sense, and I would even concede that it is the desire to transplant organs from a deceased donor patient to a recipient patient that has driven the historical progression of this construction.

But the protocols that organize these practices reveal ontological commitments that are much richer than those implied by accusations of Cartesian reductionism, and, in fact, map neatly onto the Aristotelian variations explicated in the preceding section. Close attention to the practices involved in brain-dead donations reveal at least three distinct *téchnai* which must be, and are carefully managed and mediated. The brain-dead body as recently departed donor patient is situated within the life-saving *technê* of the trauma team and the grieving practices undertaken by the family (often with help from some manner of professional grief counselors). The brain-dead body as potential body parts for transplant recipients is presided over primarily by transplant surgeons. And these two contexts are kept distinct from one another by a third mediating *technê* which presides over the brain-dead body as an actively living body that is not a patient subject—that of organ procurement organizations.

These three praxical contexts were not always so distinctly articulated. Prior to the passage of the National Organ Transplant Act (NOTA) in 1984, the status of jurisdiction over deceased donor bodies in the United States was a legal gray zone. The surviving family's rights to deceased bodies were treated as a quasi-right, extending only to the burial/disposal of the corpse, without any clear articulation of rights for the transfer of the body or its organs to other parties. NOTA was written and passed in response to concerns over an emerging commercial market for organs outside of the hospital setting (Mayes, 2003). Among other provisions, this act established a National Organ Procurement Network as a registry for organ matching, and authorized the US Department of Health and Human Services to facilitate the establishment of Organ Procurement Organizations (OPOs) (NOTA, 1984).

OPOs are not simply charged with the logistics of matching donor organs to recipients. According to the Organ Procurement and Transplant Network's policy guidelines, OPOs are responsible for diverse tasks including: obtaining formal consent for donation, clinical management of the deceased donor body, the assessment of donor organ quality, and often the recovery, preservation, and transportation of organs (OPTN Policies, 2014, Section 2.2). There are currently fifty-eight registered OPOs in the United States,[14] and while exact policies and protocols are worked out between hospitals and their local OPOs and thus vary from hospital to hospital, there is some relative homogeneity across the board.

Perhaps the most notable aspect of nearly all brain-death donation protocols is that potential conflicts of interest are scrupulously policed by way of maintaining clear boundaries between trauma staff, OPOs, and transplant surgeons. For example, only the trauma staff may declare a patient to be brain dead (after which point, the body is no longer referred to as a "patient," but as a "donor body") and OPOs are not permitted any access to the family or the donor body until this has occurred. In cases of brain-death, the opportunity for family members to spend time with the body and recover any mementoes (*e.g.* locks of hair) is made available.[15,16] While grief counseling is not done by OPOs, they typically liaise with the appropriate personnel to ensure that the donor families have the necessary support when they leave the hospital. And though there is recent discussion over the best ways and times to begin the discussion of the possibility donation with family,[17] in many hospitals *only* the OPO specialists are permitted to broach the subject. Only after the declaration of brain-death, the official authorization for donation, and the opportunity for family members to spend time with the deceased body, may organ procurement staff begin their work.

Furthermore, the OPOs are not merely an extension of the praxical context presided over by transplant surgeons. A single donor body can provide life-saving organs for up to eight recipients (this does not include "non-vital" eye and tissue donation), and this puts the OPOs in a mediating role between multiple prospective transplant recipients. They are charged with assessing the viability of potentially donatable organs, and must apply matching criteria by way of a point system in order to determine which candidates receive organ offers.[18] OPOs are entitled to withdraw offers in the event that a transplant hospital is unable to make acceptable arrangements for receiving organs (OPTN Policies, 2014, Section 5).

That there are protocols in place to carefully mediate between the distinct contexts of body as absent deceased patient, body as active non-patient brain-dead body, and body as potential organs for particular transplant recipients, is clear enough for anyone who peruses policy guidelines or discusses the matter with the professional staff involved. A brain-dead body is at least three different objects throughout the donation process, and as authority is transferred from one to the other, a different group of professionals presides over each. Furthermore, it is not a stretch to note that in working on three distinct objects in three distinct praxical contexts, trauma staff, OPOs, and transplant surgeons are engaging in distinct medical *technai*. In many instances, these distinct *technai* are even mutually antagonistic. Many aggressive life-saving interventions on the part of the trauma team (e.g., the administration of norepinephrine to combat low blood pressure—a common issue in trauma patients) can result in damage to organs[19] that either renders them unfit for transplant, or which must be actively undone by OPOs after they have received the donor body.[20] The work of clinical care of a non-patient, brain-dead body is a peculiar kind of medical practice, with goals, techniques, and accreditations that are distinct from other medical practices.

## MEDIATING MULTIPLICITIES

With the establishment of OPOs as fulfilling a key mediating role in organ donation practices, we can see that the three brain-dead bodies explicated in our variational analysis are also represented in and structure three distinct praxical contexts within the process of organ donation. Thus, organ donation ought to be viewed as involving conceptions of bodies that are admittedly peculiar, but far richer and more textured than what is suggested by accusations of Cartesian/neo-Cartesian reductionism. I should take care to acknowledge that these three variations do not fully exhaust all of the multi-stabilities available. Within the conception of brain-dead body as deceased patient, a wide variety of religious/spiritual commitments and grieving practices yield a wide variety of "absent" patients. Within the conception of brain-dead body as a living non-patient, the variable conditions of that body and its organs yield a wide variety of possible donor bodies. Depending on the satisfaction of matching criteria and organ demand, we may have a wide variety of brain-dead bodies as potential organs for recipient patients. Nonetheless, I would argue that all of these further variations can be classified as variations on the three major categories I have articulated.

In order to ensure a fluid and functioning donation process, these praxical contexts must be effectively mediated. Each moment in the progression from deceased patient, to non-patient body, to recipient patient, must be afforded adequate space and time to play itself out. This mediation through effective protocol is still evolving[21] and varies from hospital to hospital, and there are serious concerns regarding how to balance competing desiderata such as increasing available donor organs, ensuring the long-term health of transplant recipients, and respecting the needs of grieving families. It is with an eye toward managing these multiplicities that it would seem that a postphenomenological sensitivity is useful. Don Ihde has suggested that postphenomenologists have a role to play as "science critics" (Ihde, 1997), but I would extend this suggestion. The ability to freely shift between contextual gestalts and to recognize and manage multistabilities also has a crucial role to play in designing policies that deal with multiple, potentially antagonistic praxical contexts, particularly ones involving the sort of ontological multiplicities found in organ donation.

## NOTES

1. http://optn.transplant.hrsa.gov/data/
2. Theodor Kocher's successful transplant of thyroid tissue to restore lost function following thyroidectomy in 1883 (Schlich, 2010, 31–46).
3. While there is merit to arguments that the received "Cartesian" view doesn't quite square with the subtleties of Descartes's actual views of the body (as evidenced in *Passions of the Soul* or his correspondences with Mersenne and Elisabeth of Bohemia), this is beyond the

scope of this investigation. The term "Cartesianism" in this paper should be understood to refer to the received Cartesian view, regardless of whether or not it was what Descartes actually intended.

4. *cf.* Herbert Marcuse's *One Dimensional Man* (1964).

5. *cf.* Martin Heidegger's "The Question Concerning Technology" (1954).

6. I take the medical fact of brain death and its reliable diagnosis to be two related but distinct issues. For more, see Wijdicks, 2002.

7. "Form" here is not the mere "shape" or "outward appearance" captured by the term *eîdos*, but is the significantly richer term *morphé*.

8. These complications are not insurmountable, and resources for thinking of "cyborg" entities can be found within Ihde's work (e.g., Ihde, 2008,) as well as thinkers such as Donna Haraway (1991).

9. *Metaphysics* Z11, 1036b, *de Partibus Animalium* 1.1, 640b–641a, *Politics* I, 1253a, *cf.* also Lewis, 2013, 180–181, and Ackrill, 1972.

10. This may strike us as a bit strange. After all, in calling it a "severed hand" it seems as if we are identifying this flesh and bone as a special case of an ordinary hand. Additionally, we would not say the same things about, say, an eye that is shut and therefore not performing the essential function of an eye—that it is merely flesh. Nor would we say of a chariot wheel that has been removed from a chariot that it is merely wood and metal because it neither participates in the essential relationships to the rest of the chariot parts nor performs the essential function of a chariot wheel. But a shut eye can simply be opened and do all the things that eyes do, and the chariot wheel can be reattached. The shut eye and detached chariot wheel may not be *presently* active as an eye or chariot wheel, but their respective flesh, wood and metal are *potentially* active.

11. It is, additionally, worth pointing out that this intending of parts of the body as potentially parts of a whole recipient body need not wait for the pronouncement of brain death. It is already at work when a living subject agrees to be an organ donor, and even when she is simply considering such a decision.

12. *Cf.* also *Metaphysics*, 1022b (Aristotle, 1952).

13. Heidegger's articulation of this sort of "presencing through absence" may be particularly familiar to philosophers of technology due to his elaboration of the idea in *Being and Time* in his discussion of tool use, and a tool's becoming "present-at-hand" when it "breaks down."

14. organdonor.gov

15. While this is a common practice in the US, it is official policy in the UK (Donor Family Care Policy, 2004).

16. "Cardiac death" (sometimes called "circulatory death") donations are far more complicated and time sensitive, as they require that life-support be removed before waiting for cardiac arrest (which does not always promptly occur). As soon as cardiac arrest does occur, the OPOs must immediately begin the recovery of organs. Because of these and other complications, cardiac death donation is far more rare than brain-death donation. Nonetheless, OPTN Policy guidelines still stipulate that no organ recovery staff shall be present for the withdrawal of life-support in cases of donation after cardiac death. This ought to serve as a testament to the efforts made to maintain clear boundaries between praxical contexts.

17. In 2011, the United Network for Organ Sharing (UNOS) altered its guidelines, recommending earlier discussions in cases that are candidates for "cardiac-death" donation. For discussion of this decision see "When the Family Pushes and 'Decoupling' is Challenging" 2011, "Timing Can Be Everything In Organ Donation" 2011, and "Concerns Over New Organ Donor Guidelines Overblown" 2011.

18. Occasionally, transplant surgeons will additionally travel to personally assess organ quality themselves before deciding whether or not to accept an offer.

19. *cf.* Bellomo and Giantomasso, 2001.

20. *cf.* McKeown, Bonser, and Kellum, 2012.

21. This evolution may well change dramatically as new techniques and technologies emerge. In particular, the prospect of "homegrown" organs created from stemcells represents a rapidly developing option on the horizon that will surely restructure the nature of organ transplant. But it does not look as if transplants from deceased donors are going to go away anytime

soon. For more on the anticipated relationships between allotransplant medicine and regenerative medicine, see Orlando et al., 2013.

## REFERENCES


Ackrill, J. L. (1972). "Aristotle's Definitions of *Psuche.*" *Proceedings of the Aristotelian Society*. 73: 119–133.

Aristotle. (1992). *De Partibus Animalium I and De Generatione Animalium*, transl. D. M. Balme. Oxford: Oxford Univ. Press.

Aristotle. (1952). *Metaphysics,* transl. Richard Hope. Ann Arbor: Univ. of Michigan Press.

Aristotle. (2001). *On the Soul and on Memory and Recollection,* transl. Joe Sachs. Santa Fe: Green Lion Press.

Aristotle. (2001). *Physics*, transl. Joe Sachs. New Brunswick: Rutgers Univ. Press.

Aristotle. (2000). *Politics,* transl. Benjamin Jowett. Dover Publications.

"Concerns Over New Organ Donor Guidelines Overblown." (2011). retrieved from: http://abcnews.go.com/blogs/health/2011/09/20/concerns-over-new-organ-donor-guidelines-overblown/ (3/28/14).

"Donor Family Care Policy." (2004). Published by NHS: UK Transplant. Retrieved from: http://www.organdonation.nhs.uk/about_transplants/donor_care/donor_family_care_policy/donor_family_care_policy_version2_23032005.pdf (3/28/14)

Hacking, I. (2007). "Our Neo-Cartesian Bodies in Parts." *Critical Inquiry*. 34(1): 78–105.

Haraway, D. (1991). "A Cyborg Manifesto: Science, Technology, and Socialist-Feminsism in the Late Twentieth Century." In *Simians, Cyborgs, and Women: The Reinvention of Nature*. New York: Routledge.

Heidegger, M. (1939). "On the Essence and Concept of Φυσις in Aristotle's *Physics* B, I." In W. McNeill (ed.), *Pathmarks*. Cambridge: Cambridge Univ. Press.

Heidegger, M. (1954). "The Question Concerning Technology." In D. Krell (ed.), *Martin Heidegger: Basic Writings*. San Francisco: Harper.

Ihde, D. (1997). "Why Not Science Critics?" *International Studies in Philosophy.* 29(1): 45–54.

Ihde, D. (2002). *Bodies in Technology*. Minneapolis: University of Minnesota Press.

Ihde, D. (2008). "Aging: I Don't Want To Be a Cyborg!" *Phenomenology and the Cognitive Sciences*. 7(3): 397–404.

Ihde, D. (2009). "What is Postphenomenology?" In *Postphenomenology and Technoscience: the Peking University Lectures.* Albany: SUNY Press.

Leder, D. (1992). "A Tale of Two Bodies: The Cartesian Corpse and the Lived Body." *Philosophy and Medicine*. 43: 17–35.

Leder, D. (2002). "Whose Body? What Body? The Metaphysics of Organ Transplantation." *Philosophy and Medicine*. 60: 233–264.

Lewis, F. A. (2013). "Form and Matter." In G. Anagnostopoulos (ed.), *A Companion to Aristotle.* Wiley-Blackwell, 162–185.

Lock, M. (1997). "Displacing Suffering: The Reconstruction of Death in North America and Japan." In A. C. G. M. Robben (ed.), *Death Mourning and Burial: A Cross-Cultural Reader*. Oxford: Blackwell Publishing.

Lock, M. (2002). *Twice Dead: Organ Transplants and the Reinvention of Death*. Los Angeles: University of California Press.

Marcuse, H. (1964). *One Dimensional Man.* New York: Routledge.

Mayes, G. (2003). "Buying and Selling Organs for Transplantation in the United States." *Medscape Education.* 4(3): 1–4.

McKeown, D. W., R. S. Bonser, and J. A. Kellum. (2012). "Management of the Heartbeating Brain-Dead Organ Donor." *British Journal of Anaesthesia*. 108(1): 96–107.

NOTA (National Organ Transplantation Act). (1984). Pub L. 98–507, 98 Stat. 2339–2348 (October 19, 1984).

OPTN (Organ Procurement and Transplantation Network) Policies. (effective 3/7/14) retrieved from: http://optn.transplant.hrsa.gov/ContentDocuments/OPTN_Policies.pdf 3/28/14.

Orlando, G., S. Soker, R. J. Stratta, and A. Atala. (2013). "Will Regenerative Medicine Replace Transplantation?" *Cold Spring Harbor Perspect Med.* 3(8): 1–14.

Schlich, T. (2010). *The Origins of Organ Transplantation: Surgery and Laboratory Science, 1880s–1930s*. Rochester, NY: The University of Rochester Press.

Sharp, L. (2006). *Strange Harvest: Organ Transplants, Denatured Bodies, and the Transformed Self.* Los Angeles: University of California Press, Los Angeles.

Sunstein, C., and R. Thaler. (2009). *Nudge: Improving Decisions about Health, Wealth, and Happiness.* NY: Penguin Books, 177–184.

"Timing Can be Everything in Organ Donation." (2011). retrieved from: http://abcnews.go.com/blogs/health/2011/10/07/timing-can-be-everything-in-organ-donation/ 3/28/14.

"When the Family Pushes and 'Decoupling' is Challenging." (2011). retrieved from: http://www.caringfordonorfamilies.com/ (3/28/14).

Wijdicks, E. F. (2002). "Brain Death Worldwide: Accepted Fact, But No Global Consensus in Diagnostic Criteria." *Neurology* 58 (1): 20–25.

## *Chapter Thirteen*

# Towards a Hermeneutics of Unveiling

Jan Kyrre Berg Friis

Edmund Husserl's phenomenology is an introspective analysis of the intended and the intending in experience. The main purpose of phenomenology, in Husserl's view, is to return to the fundaments of knowledge. Husserl believed he could "bracket" all empirical and metaphysical assumptions, and through other methodical stages, describe essential structures of both the experienced thing and the experiencing mind with the intention to constitute knowledge that is free from noise and bias by individual and social experiences.

In Don Ihde's *Expanding Hermeneutics* (1998), we find a quite different attitude towards experience. The time that has passed between Husserl's writing during the turn of the twentieth century, and Ihde, there has passed a century of extreme change that has transformed all aspects social, technological, scientific, and philosophical. The phenomenological introspective method has been replaced by a new method of hermeneutics inspired by Heidegger and Gadamer—and in Ihde's case also by American pragmatics and philosophy of technology of the continental philosophy tradition. This implies a changed notion of experience altogether. Ihde states that experience, in this modern age of technologies, transcends the conscious subjectivity of the singular person. Technologies have become extended sense organs, enabling perceptions of realities never before known to man. The information obtained through technological mediations is not "pure." There are several problems attached to the production of technological mediations and the interpretation of the information contained in them. Technologies are detection devices, initially the input is collected, for instance, at a certain nuclear energy level, there after the technology has to "translate" the input to an energy level that is perceivable by man. Then the observer interprets and describes the information contained in the image and eventually communicates the information to other people. One of the many problems accompanying observations of this kind is observer variations—which happens when two or more observers cannot agree about the nature of the information presented in the image. The problem of observer variations arises with the observer. With the observer we have "noise" or bias disturbing the reading. Moreover, technologies may err and produce artifacts that also add to the disruptions of the truth-to-nature purity we want of information.

In this chapter I want to discuss some of the original ideas of phenomenological hermeneutics. I intend to pursue the notion of the "unveiling" taking place when we are in the act of constituting the gestalt or the object "seen." In my optics, hermeneutics is about what constitutes experience. Gadamer provides us with an understanding of the "unveiling" which concerns the perceiving mind in its more generic role as a pre-conscious "unveiling" of meaning (Gadamer, 2007, 420).[1] In Gadamer's view, understanding is an event, something that happens and something that takes shape beyond conscious awareness—pre-consciously. Human understanding has embodied sources natural and cultural, biological and social. Understanding, in its essence, is not something we are aware of doing—understanding comes to us in sudden leaps whilst we ponder a problem, when we both tacitly and consciously are interpreting the phenomena. In Gadamer's own words, hermeneutics is "not what we do, not what we should do, but on the contrary what happens to us beyond what we consciously want and do" (Gadamer, 2007, 474). In other words, a hermeneutics about science practices, discussing the material hermeneutics of technologies—as an extension of human experience of the world beyond our perceptions, should also deal with the genetics of phenomenal or perceptual gestalts—from where understanding springs.

## RADIOLOGISTS: IN-BETWEEN TECHNOLOGICAL MEDIATIONS AND PERCEPTUAL CONUNDRUMS

There are probably just as many "systematic approaches" to reading X-rays, PET scans or MRI scans, or visualizations from cardiology or pathology, as there are specialized physicians trained to conduct these tasks. There are conducted approximately a billion imaging examinations worldwide every year. There are over five million hits on Google for systematic of x-ray reading, and 35 million hits for interpretation of X-rays, just to give you a number to highlight the interest for improving medical image perceptions. Literature on "medical imaging" and on "interpretation and reporting," are plenty, all are struggling with the problem of interpretation practices having very few standards. Moreover, reports that are sent by the radiologist to the clinician are often misunderstood or misinterpreted due to faulty communication—there are no rules for communicating findings either. There are, in other words, a vast number of systematic approaches and practices that may work for the single image reader reading the image. However, when reporting the findings this personal systematic may complicate reporting, for instance it is quite typical, as radiologist Elizabeth Krupinski writes, that "decisions are not always absolutely conclusive, [they] are often formulated with plausible alternatives, and errors in interpretation can and do occur regularly" (Krupinski, 2011).

Radiological research has so far dealt with this problem by focusing on two ways of reducing diagnostic error. The first has to do with the strong belief that there is a technological "fix" to observer related problems. The emphasis is on improving technologies so that abnormalities become more visible. Viewing conditions can be optimized: better luminosity and better monitor resolution can easily be achieved. There are also efforts to improve diagnostic tools like CAD (an automated image interpretation tool). Good technologies are necessary, but not enough. What the other approach emphasizes is the need to understand how the radiologist *interacts* with technological mediations, that is, with the information in images, during the reading process. As Manning, Gale, and Krupinski write, "It is important that development and availability of CAD do not detract from the quality and need for radiological skills" (Manning, Gale, and Krupinski, 2005, 683–685). And a crucial skill is visual perception.

## RADIOLOGY AND PERCEPTION

Image interpretation consists of two fundamental processes—looking at the images (visual perception) and rendering an interpretation (cognition) (Krupinski, 2011, 393).

Here are a couple of particularly solution-resistant problems regarding perceptions in radiology. According to radiologist Harold L. Kundel, radiologists regard image analysis as the primary task, and have assumed that their perceptions of images represent the content of the images (Kundel, 2006, 402). Radiologists are usually not concerned with the process of perception itself, until it fails. Errors may occur because there are many ways that images are produced and displayed (Krupinski, 2011, 393). Krupinski says that "detecting and recognizing lesions on plain film radiography or planar scintigraphy is difficult because anatomic structures overlap, and the radiologist must translate a 2-D image into a 3-D mental representation to disembed and localize structures and lesions" (Ibid). The point I am trying to make is that overlapping structures have a camouflaging effect. It is not easy to spot abnormalities that grow around existing tissue that thus have features similar to the exiting anatomy. CT, MRI, and PET can be used to overcome this problem by producing multiple slices through the anatomy, however, the radiologist still has to search through the information to detect the abnormalities (Ibid). Medical images need to be interpreted. Interpretations ultimately rest on human perceptions and skill to interpret.

At least half of the errors made in clinical practice are perceptual (Krupinski, Kundel, Judy, and Nodine, 1998, 611–612). For example, the radiologist is faced with a lesion and he knows the location only approximately. He is told to look for a lung nodule. The radiologist will here always apply his experience and knowledge; or rather he knows where nodules are likely to hide in the lungs (ibid.).[2] Such peripherally inconspicuous targets have to be found by scanning the area. Experiments with eye-trackers show that the search pattern of the observing radiologist is influenced by both clinical history and experience (Kundel, 1990, 472–483). This is noteworthy since most of us already know of the impact of context, training, and experience from phenomenology, hermeneutics, and the philosophy of perception. An example of being influenced by clinical history is from one of my visits at the Radiology Department at Bispebjerg hospital here in Copenhagen—they have generously granted me permission to conduct observations and interviews with radiologists and technicians working there. A few years ago an elderly lady had fallen and fractured her femur, the fracture was complicated and she had to undergo surgery several times to fixate and stabilize the fractured area. During this process a number of X-rays were taken and stored on to the PACS (Picture Archiving and Communication System). One year after the first hospitalization she had a follow-up at the hospital, and a new set of eight images were taken: seven images of the lesion area and another showing her whole torso down to her knees in profile. In August this year she was hospitalized again—this time with a huge cancer tumor in her stomach. During morning conference in the end of August, the radiologist responsible for describing her X-rays got the images taken at her follow-up, a couple of years prior, up on the right screen of the PACS to compare. Everyone present in the room could clearly see the cancer on the years-old torso image, and they had to admit that this was something they had overlooked, simply because they went straight into the trap and focused solely on the lesion pointed out to them in the requisition. This is a common error and has been termed "satisfaction of search." As soon as the radiologist has found his fracture he is satisfied with both having identified and located what he was searching for. The problem is that one may have not perceived everything and have thus abolished the search prematurely.

Radiological practice invariably brings with it observer variations. Observer variations are complex. This is a seemingly unsolvable epistemological problem. For instance, computers have become immensely important in displaying image information. For radiologists this means that the information must be perceptible and as such be capable of being extracted. The radiologist has an imaging task to do. This task defines what information is needed in order to come up with an interpretation leading to a diagnosis. Here the quality of the image is of course very important. Kundel writes:

> Visualization and estimation tasks may require different visual cues for their performance. Visualization tasks require strong boundary cues, whereas intensity estimation tasks require texture cues. Both tasks can be aided by appropriate image displays. Color can be helpful in estimation tasks, and three-dimensional display can aid visualization tasks. No matter what the task or the image is, as long as an observer is needed to read out the information, performance is the final arbiter of the goodness of the image. (Kundel, 1990, 472–83)

In other words, interpretative efforts on the radiologist's part depend heavily on the technology's translational functionality. Following Don Ihde, the radiologist's interpretations are made "easier," which is, less challenged by the technology through the interpretative functionality of the technology itself. The technology "embodies" a "hermeneutic" (Ihde, 2009, 56). MRI and CT, x-ray and ultrasound, are all translational technologies that not only translate or transform a phenomenon into a readable image but add color and other graphics in order to enhance the readability of the phenomenon in question (Ihde, 2009, 56). Ihde calls the technological translation of the real world phenomenon into a mediation or image a "material hermeneutic process" and brings medical image interpretation in somewhere between the natural and the humanistic science practices (Ihde, 2009, 64). Another perspective on the radiologist's practice as image reader is from cognitive research on perception. Currently various methods for assessing diagnostic image quality are used. These methods are usually evaluations based on individual clinical experience, measurement of diagnostic performance, and physical measurements made on images or imaging systems. The reason many use psychophysics to investigate observer variability is its assumed inherent promise of a stable solution, which the development of a certain mathematical model will allow for stable predictions of the system output from any arbitrary input (Kundel, 1979, 265–271).

Psychological work on perception has largely but not exclusively taken place within the theoretical framework of "cognitivism." Within cognitivism the application of psychophysical methods to map brain activity has been the traditional scientific approach. Cognitive psychologists are particularly interested in the quantitative relationship between a visual stimulus and the perceiver's response. The basic idea behind cognitivism is that all intelligence resembles computation and cognition can thus be defined as "computations of symbolic representations" (Varela et al., 1993, 40). The problem with cognitivism in relation to experience is that cognitivism postulates mental processes that never are, or perhaps cannot be, brought to the level of consciousness. Indeed there are pre-conscious processes that are automated.

However, in order for the processes to become automated or embodied, they have to be brought to consciousness in order to function properly (Varela et al., 1993, 49). In other words, there have to be a link between intentional experience of the perceiver and embodied automated skills framing the perceptions with meanings. According to Kundel imaging scientists would love to predict how an observer will respond to any image configuration without having to bother with the messy business of performing a study with real human observers (Kundel, 2006, 402–408). However, this may become impossible by route of cognitivism and its disconnectedness or fragmentation of preconscious and conscious mental states. This has led Kundel to call for "research in the deeper aspects of image perception and in the interface between perception and analysis" (Ibid). I take this to mean a deeper understanding of the hermeneutics of medical image perception.

## HERMENEUTICS IN MEDICINE

Hermeneutics is embedded into medical "image-interpretation-practices," as it is into all science practice. Modern hermeneutics engages in all human activity and the products of this activity, and in this chapter, with the *interpretation* of images. Thus, hermeneutics is a method unknowingly applied by radiologists in describing images from the whole range of medically relevant mediations.[3] Medicine is as a scientific field permeated with technologies, especially technologies designed to visualize real world physical phenomena unreachable by the human senses. Radiologists are of course aware of their interpretative participation in the reading of images, but their explanatory framework has no reference whatsoever to hermeneutics as a specific way of systematizing one's interpretive efforts. Neither is there any talk of how hermeneutics is imbedded in the act of perceiving, that observers already have some skills to make sense of what they see. In radiology they don't call it "hermeneutics" they call it "information-processing theory" and it constitutes the theoretical basis for interpretations of visual search data. The interpretation processes have several stages and hermeneutics and information-processing theory overlap almost exactly.

Helmholtz stated that eyes have poor optics. He said that eyes are organs that are elegantly designed for daytime hunting, for rapidly obtaining information of large objects, but no design for detailed analysis (Helmholtz quoted from Sabih et al., 2010, 1–2). Besides the biological and neurophysiological programming of our visual system—and the physical aspects of the image—there also are the psychological and emotional factors influencing the way all of us perceive pictures, and in particular how radiologists perceive medical images. These processes are of the form of pattern recognition, spatial localization of the part of interest and the comparison of patterns with known patterns—this is also known as Gestalt processing (Koontz and Gunderman, 2008, 1156–1160).

We may therefore claim that interpretation—or extraction of details from complex image backgrounds, relies on two processes, both oriented towards the visual field or the image. The primary mode of perception is pre-conscious, automatic, rapid and tacit. The eyes are always moving about, called saccades, which scan the image and bring different areas to the fovea. The input at this stage is discontinuous and jerky, with a lot of noise and blurred images being projected during movements (eyes move up to 400 degrees per second). We should keep in mind that 10 billion bits of information arrive at the retina, 6 million bits enter into the optic nerves, and only 100 bits per second constitute conscious perception. The brain will at this stage often jump to conclusions that might be erroneous.

The other mode is a reflexive, problem solving, slow, sequential, effortful modus, which is difficult to sustain (Sabih et al., 2010, 1–2). Actually there are three levels of performance. To be able to read an image we need skills and skills are patterns of thought and actions governed by a certain schemata or pre-programming. The literature, for example, Sabih et al. (2010, 1–2) or Rasmussen and Jensen (1974, 293–307), claim that the second aspect is the rule based processing, where solutions of familiar problems are stored as rules of the type "if x then y." The third is rather obvious and is the knowledge based aspect, here analytic processing is applied to synthesize known patterns and to derive inference about new findings.

The initial look at the image results in a global impression—where the radiologist extracts information about anatomy, color, symmetry, and gray-scale content. This information is then compared with information stored in long-term memory—it is this stored information, which belongs to the observer's horizon or life-world—that forms the viewer's expectations of what kind of information may be contained in the image. Something can be detected at this stage, but since medical images are very complex, this is rarely the case. To identify abnormalities focal search is required, which is an interactive process partly directed by the viewer's expectations or cognitive schemata. Features are examined closely and compared with stored experience and attained knowledge functioning as part of the cognitive schemata tacitly applied in the reading process. When image features match with the cognitive schemata a decision is made. Throughout the search the observer has to cycle between global overview and focal search modes, between whole and part. So much for the process of interpreting images, it is also important to bring in the communication of findings, since these rests on the radiologist's choice of systematics.

## COMMUNICATING FINDINGS OR REPORTING READINGS

There are other challenges to grapple with, and with which the radiologist is typically confronted every day. An integral part of the radiologist's praxis is to report his visual impressions to the clinician. As I have tried to show—the whole "perception—cognition—interpretation—transmission—understanding" process is very complex (Sabih et al., 2010, 1)—and it is therefore no wonder that errors might occur somewhere in the chain (Ibid). This problem was particularly pointed out to me during interviews with nuclear radiologists at the Radiology Department at Bispebjerg hospital in Copenhagen. These interviews revealed a fundamental lack of standardized and structured reporting, and moreover, that the communication between the radiologist and the clinician is colored by jargon due to sub-specialization of the radiologist. Recent literature has also begun investigating the communication line. For instance, it is said that "during radiology training, too little attention is given to structured report writing skills, and trainees have to *hone their own skills by learning from the varied methods of different senior colleagues*" (Chahatani, Sahu, and Sankaye, 2012, 722–725). Other studies report of a wide use of unfamiliar and undefined terms (Espeland and Baerheim, 2007, 15–19). And again others state that "There is a wide variation in the language used to describe imaging findings and diagnostic certainty" (Pool and Goergen, 2010, 634–643).

The whole effort of improving technology—whether it is the automated diagnostic tool (CAD) or image technologies themselves, is of no use as long as the communication of what is found is not understood by the clinician. Again there are socio-psychological trapdoors here as well. There is a *pressure to report*—meaning that there is a certain "need" to find something wrong with the patient, so findings, often insignificant are reported in a language that is ambiguous and might be misinterpreted as something significant (Sabih et al., 2010, 5). In other words, there is a certain need to improve the communication between radiologist and clinician; one such improvement could be to standardize the language used. However, it is the interpretation practice that lurks in the background. By that I mean to say that radiologists interpret and in order to do so, apply their own systematic when they read. Therefore there is no wonder that reports are structured differently. The aspect of sub-specialization is also important. Some years ago, it was more natural among radiologists to be a generalist, that is, being able to read images from different imaging technologies, say, both conventional X-rays and MRI scans. Today, according to the radiologists at Bispebjerg hospital, it is usual to specialize and to handle one product alone. According to the same physicians, some sort of sub-culture has developed as a result of this and one consequence is the emergence of a specific jargon or sub-language thus used to describe findings. So there is a connection between the systematics of interpretation, sub-specialization and applied language.

## TOWARDS A HERMENEUTICS OF UNVEILING

Hermeneutics is an unveiling of the ontological conditions for understanding. Hermeneutics is about understanding the interpretative mechanism, and whatever is influencing the way we understand, and thus also what we understand. How we understand, is something we can return to, reflexively. The "how" is something we all have access to intuitively. If we consider the external influences—and the "noise" embodied, we need to access all kinds of scientific resources of possible influence in order to identify what we have embodied. Investigating experience is essential, but, alone, it is not enough. Philosophy need to become experimental.

The new insights into the duality of moral intentionality are informative. Experimental philosophy has shown that *how* we actually act *when we have to* stands in stark contrast to how we think we will act if the situation should occur. History tells us, and deep down we have to admit the truth of it, self-preservation and self-interest are always close to heart—given the right circumstances we will attempt to "preserve" ourselves. Human nature seems to be absorbed in the "here-and-now" of individual existence—and always will be, it is human nature, and it is an essential characteristic of how the mind works. We are forced to reckon with the present and to act according to what happens *when it happens*. We are always dealing with life in its ever-changing and present modus. We are always relating to concrete realities in the present. Never do we get the opportunity to escape the actuality of our situation.

Human morality is context dependent. And human interpretation in its generic mode is about context—before consciousness. As a meaning-generating *act* it is about *reacting* according to context. The reacting to the actual includes a creation or a leap forward from past experiences to that which might happen, but which has not happened yet. The reason for this is that reactions do not always tell the truth about what actually happens. There is a temporal aspect we cannot go beyond. We have to wait for the unfolding of the event itself; however, we don't have to. In acute situations we never wait for the outcome. We actually react on incomplete information, on gestalt-information before we truly understand what is going on, information offered by past experiences and a creative mind operating tacitly and pre-consciously. When that is said, we obviously do manage to break free in order to explore and go new directions; to conquer a habit that holds us back and makes us take the wrong decisions. We have the ability to interpret before awareness on fragmentary gestalts and loads of embodied know-how. We do occasionally break free and follow the intuition which throws us on to a more insightful level of understanding.

It is not enough to approach the complex nature of interpretation from a re-reading of the classics—philosophers of science, postphenomenologists, technoscience theorists, all need to immerse themselves in real empirical study, to study what the scientist do. By immersing themselves in the culture and practices of interpretation familiar to the scientist, they are opening up a passage to themselves into the scientists' domain of understanding.

## NOTES

1. I follow Hans-Georg Gadamer in his application of Heidegger's interpretation of the Greek word for truth, *alétheia*, meaning undisclosed or unveiled, which is to mean an unveiling of that which has been disclosed or hidden.
2. The example is borrowed from the Krupinski paper.
3. Patrick Heelan and Don Ihde are the two of several known exponents of the hermeneutical approach within the natural sciences.

## REFERENCES

Beyer, C. (Winter 2013 Edition). "Edmund Husserl." *The Stanford Encyclopedia of Philosophy*, Edward N. Zalta (ed.), http://plato.stanford.edu/archives/win2013/entries/husserl/.

Chahatani, S., A. Sahu, and P. Sankaye. (2012). Reply to the paper "The Radiology Report—Are We Getting the Message Across?" *Clinical Radiology* 67: 722–725.

Espeland, A., and A. Baerheim. (2007). "General Practitioners' Views on Radiology Reports of Plain Radiology for Back Pain." *Scandinavian Journal of Primary Health Care* 25: 15–19.

Gadamer, H.-G. (2007). *Sandhed og Metode*, Danish translation by Arne Jørgensen of *Warheit und Metode* (1960–1990).

Ihde, D. (2009). *Postphenomenology and Technoscience: The Peking Lectures*. Albany: SUNY Press.

Ihde, D. (1998). *Expanding Hermeneutics. Visualism in Science*. Evanston, IL: Northwestern University Press.

Ihde, D. (2012). *Experimental Phenomenology: Multistabilities*, 2nd edition. Albany: SUNY Press.

Koontz, N. A., and R. B. Gunderman. (2008). "Gestalt Theory: Implications for Radiology Education." *AJR Am J Roentgenol.* 190(5): 1156–1160.

Krupinski, E.A. (2011). "The Role of Perception in Imaging: Past and Future." *Seminars in Nuclear Medicine* 41(6): 392–400.

Krupinski, E. A., H. L. Kundel, P. F. Judy, and C. F. Nodine. (1998). "Key Issues for Image Perception Research." *Radiology* 209: 611–612.

Kundel, H. L. (1990). "Visual Cues in the Interpretation of Medical Images." *Journal of Clinical Neurophysiology.* October 7 (4): 472–483.

Kundel, H. L. (1979). "Images, Image Quality and Observer Performance." *Radiology* 132: 265–271.

Kundel, H. L. (2006). "History of Research in Medical Image Perception." *American College of Radiology.* 3: 402–408.

Manning, D. J., A. Gale, and E. A. Krupinski. (2005). "Perception Research in Medical Imaging." *British Journal of Radiology* 78: 683–685.

Pool, F., and S. Goergen. (2010). "Quality of the Written Radiology Report: A Review of the Literature." *Journal of the American College of Radiology.* 7: 634–643.

Rasmussen, J. and A. Jensen. (1974). "Mental Procedures in Real-Life Tasks: A Case Study of Electronic Trouble Shooting." *Ergometrics* 17 (3): 293–307.

Sabih et al. (2010). "Image Perception and Interpretation of Abnormalities: Can We Believe Our Eyes? Can We Do Something About It?" *European Society of Radiology* (2010): 1–2.

Varela, F. J., E. Thompson., and E. Rosch. (1993). *The Embodied Mind: Cognitive Science and Human Experience.* Cambridge: The MIT Press.

# 4

# Critical Interlocutors

*Chapter Fourteen*

# Making the Gestalt Switch

Andrew Feenberg

Don Ihde's *Postphenomenology and Technoscience: The Peking University Lectures* contains a good brief summary of his work from his earliest investigations of perception and embodiment to his latest studies of visual technologies (2009). I find in this book many parallels with my own work. As Ihde points out we are both influenced by phenomenology and belong to something called the "empirical turn" in philosophy of technology. These commonalities made it possible for me to comment sympathetically on his earlier book on the body some years ago at the APA. On that occasion I sought to supplement Ihde's focus on the body as subject with reflections on the body as object. I will follow a similar line here in commenting on this recent book, although to complete the picture I will also refer to his earlier book *Technology and the Lifeworld* (1990). Once again he focuses on subjectivity, this time in terms of the relation of scientific knowledge to nature, whereas I am interested in the object, nature, and more generally the social impact of the construction of nature-like objects in the social world, what Lukács called "reification."

Ihde now calls his approach "postphenomenology." This term signifies his synthesis of aspects of phenomenology with pragmatism. The outstanding innovation of both these philosophies was the break with the epistemological tradition and the move toward what Ihde calls a "nonsubjectivistic and interrelational" approach. However, pragmatism proved more radical in its turn toward practice whereas Husserl remained caught in the terminology and problematics of epistemology. Idhe takes over Husserl's notion of intentionality while marrying it to pragmatist concepts of practice and embodiment.

I should mention one more influence on Ihde's approach, although it is one he increasingly repudiates, and that is Heidegger's various analyses of technical practice and the technological worldview. Ihde borrows Heidegger's concept of "world" as a nexus of meanings enacted in practice, and his later notion of technology as the underlying basis of modernity. But he rejects Heidegger's romantic nostalgia for earlier technologies and his vague hope in salvation through the intervention of a "god," that is, something external to our technological world. Instead, Ihde turns to Science and Technology Studies (STS) for empirical approaches to particular technologies.

The core argument I find most persuasive in Ihde's work is the notion that human beings have always already left the garden of Eden for a technically mediated world of some sort. We are homo technologicus by our very nature. Technology is not something added on after the fact, like those peculiar little sweaters small dogs are sometimes outfitted with in winter. No. Technology is as natural to human beings as language and culture; its specific content is historically contingent but it will always be found wherever there are human beings.

Ihde employs a concept of culture as an overarching framework or pattern of artifacts, beliefs, and practices. Such a framework informs what he calls "macroperception" in his earlier book, *Technology and the Lifeworld* (1990). Macroperception in modern societies is deeply influenced by the scientific construction of objectivity and by scientific knowledge. This affects our "microperceptions," that is our bodily engagement with objects.

Ihde's version of the practical basis of a scientific-technical culture is the technological mediation of perception by new instruments. He emphasizes the role of the telescope and imaging technologies, as well as the introduction of new and more precise technologies of measurement. This approach is reminiscent of McLuhan, with his theories of the cultural impact of mediated perception. However, Ihde is guided by phenomenology toward finer analyses of the modulation of perception and practice by technological change.

This general theory of the relative universality of technology is complemented by an analysis of the specific meanings and developments of technologies in different cultures. Ihde focuses on modern technology and the emergence of a science and a lifeworld based on its achievements. This analysis is complicated by the fact that much of modern technology alters perception. He is especially interested in the telescope, both because of its importance in the history of science and also because of the role Galileo plays in Husserl's theory of the lifeworld. The telescope magnifies both the object seen and the bodily movements of the viewer. It enables Galileo to build a new view of the universe that extends and radicalizes the implications of the existing lifewordly practices of quantification and such innovations as artistic perspective and navigational techniques, all of which contribute to a new culture congruent with his mathematical science (Ihde, 1990, 64–65).

Instruments make modern science possible and influence our interpretation of nature, even our interpretation of our own sense experience. Just as human nature includes technology, so the perceptual lifeworld includes science. For example, when we look at the moon we look at a rock, not the heavenly body a premodern might have seen. Similarly, symptoms of illness refer us to a cause, not to a curse, and so on.

But unfortunately, our science-influenced perceptual culture has also been influenced by commercialism and masculinist ideology. Modern technologies are embedded in a cultural context that favors the most ruthless exploitation of nature. Ihde is concerned by the environmental crisis that has resulted from the last two centuries of technological progress under these conditions. He believes a clear understanding of the nature of technology can provide guidance to environmentalism in responding to the crisis. In contrast with the popular notion that change can come from a spiritual revolution, for example, he argues that the crisis can only be overcome through a "gestalt switch in sensibilities [that] will have to occur from *within* technological cultures" (Ihde, 1990, 200). Such a switch is possible because technologies do not stand alone. They are always interpreted and employed in a cultural context. The "multistability" of technology holds open the possibility of change "from within."

This point is specifically aimed at Heidegger. But in what would such a gestalt switch consist? If I have a criticism of Ihde's book, it is the absence of a concrete answer to this question. Given that modernity will not be saved by abandoning its engagement with science and technology, what is the alternative path we should be following? I want to pursue this question in terms of the deep background of contemporary thinking about technology. Heidegger most definitely plays the central role in that background.

A good deal of what we find interesting in philosophy of technology and in STS was anticipated by earlier thinkers in different language and with different emphases. I have become acutely aware of this during the last two years revising my first book on Lukács. The book was originally published over thirty years ago but in working on it I discovered that it underlies all my later work in philosophy of technology and is even relevant to STS. The book is now available with Verso under the title *The Philosophy of Praxis: Marx, Lukács, and the Frankfurt School* (2014).

Before I proceed with my argument I should perhaps remind you that in 1923 Lukács published the founding work of Western Marxism, *History and Class Consciousness.* This book introduced a Marxism strongly influenced by Hegel and free of the dogmatism of the Soviet tradition. It is still interesting today as an important development in Continental Philosophy with wide influence on twentieth-century philosophy, especially the Frankfurt School. Furthermore, as I will show briefly here, important aspects of Lukács's argument derive from sources that also influenced Heidegger. Hence there are interesting parallels between their work and especially between their critiques of science and technology.

Confronting my early presentation of Lukács's theory of reification with my later work in technology studies I became aware of the extent to which the germ of my approach to technology is contained already in that concept. The concept of reification is meant to convert nouns into verbs. The basic idea is that what we take to be substantial entities—things—are in fact continually produced and reproduced by practices. Hence thinghood is actually the appearance of a deeper practical reality. Where have we heard this before? I do not need to remind you that this is the basic thesis not only of pragmatism and Ihde's work but of social constructivism and Actor-Network Theory as well.

But the idea of reification goes much deeper. Lukács argues that it originates in Kant's concept of transcendental synthesis. Reality is not simply there to be observed, but must be constructed by the mind in terms of forms and categories that give it coherence and generate the appearance of thinghood we normally take for granted. But Kant's transcendental standpoint presupposes an individual subject, a consciousness, as the constituting agent. Lukács argues that the actual agent of the construction is social, not individual, and consists not in pure acts of the mind but in complex social practices.

However, Lukács does not develop case histories like our contemporaries. He approaches the practical basis of reification differently, in terms that derive ultimately from neo-Kantianism. It is noteworthy that this is also the source of Heidegger's speculations on science and technology. Both Lukács and Heidegger draw on the neo-Kantian concept of the construction of object domains. This concept is the basis for their critique of science and technology which impose a specific apriori conception of objectivity on the world in order to understand and master it. Once they have identified the logic of that form of objectivity, they explore its generalization as a cultural universal, no longer confined to specialized domains but shaping the lifeworld of everyone in modern societies.

For example, Heidegger explains science as constructing nature as the sort of thing that can be planned and controlled. This nature is subject to precise measurement and emerges as the object of experimental manipulation. Whatever cannot be measured and manipulated is consigned to the realm of secondary qualities. The real is now defined in terms of this construction. Heidegger identifies it with technology. Like Lukács, Heidegger argues that the agent of the construction is transindividual, not of course a social class or group, but an epoch in the history of being. The technological epoch is exemplified in modern machine technology but its spirit long precedes the industrial revolution and shapes the emerging natural sciences much earlier.

Lukács made a similar argument but he attributed the origin of the construction to the practices typical of capitalism, generalized in science and in a reified worldview. He focused on two practices in particular, commodity exchange and the operation of mechanical devices. Commodity exchange requires equivalence of dissimilar objects which in turn requires quantification for comparability. Marx analyzed this complex in his discussion of the "fetishism" of commodities, by which he meant not the love of consumption but the appearance of use value as exchange value, as price. Everywhere in capitalist society concrete goods take on the commodity form and present themselves through a quantitative determination. Just so the essence of the nature of natural science is a measurable, quantitative representation.

Similarly, the subject of deskilled industrial operations is external to the production process, an appendix of the self-acting machine. Obedience to the "law" of the machine's functioning is the condition sine qua non of human agency. Again, science and technology involve social practices congruent with capitalism, practices which share the worker's destiny: manipulation under the "law."

The reified worldview is based on the generalization of these aspects of capitalist practice in a concept of nature and a corresponding concept of subjectivity. Nature is a quantifiable entity and the human subject is an individual confronted with a world that cannot be fundamentally changed, only technically manipulated.

The modern lifeworld emerges from the concatenation of such a view and persisting elements of tradition and immediate sensory experience. In Heidegger little place remains for the marginal practices that are not assimilated to the *Gestell*, however, their continuing existence holds out a slight hope for a different dispensation in the future. Lukács argues that the reified lifeworld is fraught with contradiction since it cannot adequately mediate workers' needs. The contrast between the imposed capitalist forms and their lived experience is potentially explosive.

Now clearly, these apriori object constructions in Heidegger and Lukács are quite different from the empirical specifics favored by contemporary STS. Both these earlier thinkers seek to understand the modern epoch of science and technology rather than analyzing cases. But the cases belong to an already established modern framework and cannot be fully understood without reference to the nature of that framework. In different ways Ihde and I have attempted to synthesize the epochal approach with empirical studies, an operation that is possible because both types of analysis depend on dereification of the substantial appearances of the social constructions underlying modernity.

Ihde does not search for a general form of objectivity characteristic of modern societies. His growing skepticism about Heidegger has led to a different orientation. Instead he attempts to outline the impact of the perceptual changes made possible by new technological instruments on modern subjectivity. Ihde does agree with Heidegger and other critics of technology that Western societies exhibit a tendency toward "technological totalization."

They try to incorporate nature into culture through the extension of technologies into the body and the world on an ever increasing scale. This tendency distinguishes modernity from premodernity and explains the environmental dangers that loom over our future.

However, Ihde argues that this is not a complete description of what is happening in the world today. He sees in the development of a global technological civilization an opportunity to overcome the narrow dogmatisms and ethnocentricities of what he calls "monoculture," including our own. The move toward a "pluricultural" world is a democratic advance. Ihde associates it with post-modernity, a new phase in the development of technological civilization that opens new possibilities of critique and change. Exactly how this is supposed to work concretely I have not been able to figure out. But I think Ihde's main point is that the global interaction of cultures calls into question many Western prejudices inherited from the past which have led to the current crisis. Perhaps with the challenge to Western ethnocentrism technology can be resituated in another cultural context that privileges conservation and tolerance rather than exploitation and discrimination.

While I can agree with all of this, I have focused more on the construction of the concept of the object as a cultural phenomenon. My approach has led me to what I call the "instrumentalization theory." I attempt to explain the relation between causal and hermeneutic aspects of technologies, scientific-technical rationality and the lifeworld contexts of technology. This dual aspect approach to technology resembles Ihde's distinction between the "merely technical" and the cultural context. Like him I distance myself from Heidegger's dystopian logic without giving up entirely what I take to be his important discovery that the lifeworld is transformed by scientific-technical thinking in modern times.

The instrumentalization theory suggests an answer to the question I posed earlier about the alternative path our civilization must follow if it is to survive. I argue that what is required is not an escape from technology but rather its dereification. By this I mean enabling a more fluid interaction between rational disciplines, systems, and artifacts and the demands of the lifeworld of concrete experience. I call this interaction a mediation in a sense that is more or less Hegelian. Rationality does not exist separate from the lifeworld but is an extension of it along specific lines such as quantitative precision and deductive rigor. As Ihde points out, these extensions depend on technical practices in the lifeworld such as measurement and writing. Technological applications of rational disciplines are also dependent on the lifeworld context for their meaning and trajectory of development. This again is a point Ihde and I share. Technology does not transcend the lifeworld but rather forms a special part of it. This explains why the lifeworld can in turn "mediate" technology and other rational systems, taking advantage of their multistability to redefine them.

This conception has political implications. The conflictual interactions in which most mediation consists can only flourish in a democracy. The contestation in which the process of mediation goes on presupposes respect for basic democratic principles, human rights, and the will of the majority. This must be a "deep democracy" in which all forms of rational order, and not just law, are subject to dereification and transformation.

I interpret mediation in this sense through the Lukácsian critique of formal rationality. Technologies, bureaucracies, and markets are rational institutions that impose form on the stuff of everyday experience. They define human beings as objects in ways similar to the construction of nature by natural science. Where the forms prove unsatisfactory to the human beings whose lives are their content, the misfit evokes criticism and conflict. These situations reveal the relativity of modern formal rationality in its social deployments. Ihde points out that every amplification hides a reduction. This is most dramatically so where human lives are at stake. None of these rational systems is able to fully embrace its contents. Indeed, no such full embrace is really imaginable in a world where every advance in rationality creates new possibilities for differentiation among the members of society. The concept of democracy must be enlarged to include the continual readjustment of formal structures to the demands of human agents.

We already have significant examples of such democratic interventions in many domains. In addition to continuing class conflict, diminished in scope and intensity but by no means resolved, there are many types of protest movements, hacking of computerized systems, lawsuits, hearings, and forums, especially around environmental issues, and even lay participation in the work of scientific experimentation and technical design. Although their scope and effectiveness are still severely limited, the new types of interventions into formally rational systems enlarge the public sphere and contribute to such important new directions for society as environmental modernization. Yet they have been systematically under-estimated and ignored by political theorists. Where they are noticed at all, resistances are generally viewed not dialectically but from the one-sided perspective of the dominant. From that perspective rationality stands opposed to ignorance and disorder. But the reality is quite different.

These are essential forms of activism in a technological society. They limit the autonomy of experts and capitalist management and force them to redesign the worlds they create to represent a wider range of interests. The translated demands are assimilated by the institutions and may lead in turn to future iterations of the struggle, further contestation. This is the logic of reification and mediation and it is unsurpassable. I call it "democratic rationalization" because it reproduces rational institutions in response to pressure from below.

I believe this conception of the politics of technology conforms to Ihde's prescription according to which the "gestalt switch in sensibilities [that] will have to occur from *within* technological cultures." It suggests that the "pluriculture" not only crosses national boundaries but also the sharply drawn line between lay and expert, the human individual and the system which both makes possible individuality and constrains it. I look forward to hearing Ihde's own take, and those of the other postphenomenologists, on this amplification of their framework!

## REFERENCES

Feenberg, A. (2014). *The Philosophy of Praxis: Marx, Lukács, and the Frankfurt School.* Verso.

Ihde, D. (1990). *Technology and the Lifeworld.* Bloomington: Indiana University Press.

Ihde, D. (2009). *Postphenomenology and Technoscience: The Peking University Lectures.* Albany: SUNY Press.

Lukács, G. (1923). *History and Class Consciousness.* MIT Press, 1972.

## *Chapter Fifteen*

# Postphenomenology with an Eye to the Future

Diane P. Michelfelder

Now is a good time to be a postphenomenologist. While it is easy to see why, a suggestive image from Maurice Merleau-Ponty's *The Phenomenology of Perception* can still be helpful. In a well-known remark, he observed that the human body is "geared" into the world (1962, 250). The image seems apt when speaking of the relation between the body of work of postphenomenology to the world, and not only because the word "gear" is a technological term. Postphenomenology is intensely geared-in to the technological devices and systems of contemporary societies. Its interest in understanding the distinctively human user experience of technology and how technology works to shape the user experience meshes tightly with the "given" of new, innovative technologies that are constantly emerging and forming new user relations.

Consider this: when Don Ihde first described the set of human-technology-world relations as being ones of *embodiment, hermeneutic, alterity,* and *background*, the example he gave for *embodiment* relations was an ordinary pair of glasses (Ihde, 1990). Now, there is not only the anticipation of an unlimited public release for Google Glass in the foreseeable future, but some optometrists are already prepared to sell the Glass with frames and prescription lenses, and Glass-enabled contact lenses are in the works. Robots capable of brushing the teeth of children (Ihde, 2002, xii) may still be a techno-fantasy, but Internet-enabled toothbrushes connected to a smart phone app that will gather, and transmit back to users, data on their existing brushing habits will soon be on the market. These examples reflect how not only are we living in the midst of extraordinary innovation with respect to individual technologies, but also with respect to ways in which they are linked to one another, interact with one another, and come to be a part of our everyday experience. And, because so many of these innovations involve perception and embodiment, postphenomenology will have many profound insights to offer into how they are shaped by and in turn shape the human experience. To put matters more simply: wherever there is an app for that, postphenomenology cannot be far behind. If the lively currents of technological development help make the present a good time to be a postphenomenologist, we can expect the future to be a good time as well.

Much like technologies, though, philosophies are multistable. How to say what's next for postphenomenology? How might it "embody" its future? In the spirit of multistability, let me suggest two possible, partially overlapping and so non-exclusive, ways. The first future will sound familiar to many readers of this volume, as it points to a course already being charted by much current work in the field. The second future is arguably more adventurous and de-stabilizing, involving not so much a complete turn-around but a shift away from the empirical focus that now influences much postphenomenological work toward a more speculative direction.

Put concisely, in this second future, more attention would be given to forms of disclosure connected to the "world" element in the "human-technology-world" analytical frame that plays a decisive and illuminating role in postphenomenological investigations. While these two futures are not mutually exclusive, without attending more to "world"—meaning "the world as a whole"—postphenomenology, I want to propose, runs the risk of becoming a less vital voice within the profound conversations currently taking place in philosophical circles that focus on the technologies of everyday life.

## POSTPHENOMENOLOGICAL FUTURES: TAKE ONE

Traditionally, philosophical movements have progressed by refining core concepts and drawing distinctions that become ever more nuanced and precise. Postphenomenology is no exception in this regard. Following upon the groundbreaking conceptual scaffolding laid out by Don Ihde, others have moved postphenomenology forward by critically reflecting on and expanding this initial framework. Most significantly, Peter-Paul Verbeek has elaborated on the idea that technology "mediates" between human experience and the world, suggesting that mediation is not a simple connection between already existing subjects and objects, but is rather a co-shaping that carries with it an ethical dimension: new technologies generate new ways of disclosing the self and also new social practices which can add to or subtract from the quality of life and pose novel moral challenges and consequently novel demands on our moral responsibility (Verbeek, 2005, especially 121–145; 2011). Out of this focus on mediation, a rich grouping of questions emerges. In what ways do the formal structures of mediation as connection differ from mediation as co-shaping? How do other variables, in the shape of different kinds of mediation—language and social structures, for instance—influence technological mediation? How do these other variables intersect with the mediations of technology to shape human experience and the world?

Other postphenomenological investigations, moving along a complementary path, have zeroed in on the user experience of particular technologies as an occasion to extend and improve the conceptual "lenses" through which postphenomenology does its work. If technology can be said to co-shape user experience and the world, so can technology be said to co-shape the future of postphenomenology by providing a critical perspective on its analytic tools. For instance, as part of his body of work on the experience of cell phoning while driving, Robert Rosenberger has argued that this experience cannot be fully understood without expanding the scaffolding just mentioned to include a "field theory of technological relations" (Rosenberger, 2014). Such a critical perspective could also be developed in another way, in order to sound a critical note about just how flexible our experience can be in adapting to new technologies. To see this, let's go back to Google Glass.

A postphenomenological account of Google Glass might focus on how each of the two elements that go to make up this particular wearable computer are associated with different analytical "tools." Our relations to the "Glass" side of this technology would seem to be one of embodiment, particularly when coupled with a conventional pair of glasses. Things become considerably more complex once human-technology relations are seen from this device's Google side. Take the planned "wayfinding" feature of the Google side of the Glass. With this feature in place, drivers wearing the Glass would no longer need to use their vehicular GPS navigational systems or GPS smart phone applications. "Wayfinding" through Glass could in some sense be said to be safer than other means of gathering GPS navigational data, as the user would not need to take her eyes off the road in order to refer to it to chart a course to her destination. That course would be displayed iconically on the Glass itself, perhaps by means of a colored line showing the route to be followed. This particular feature of the Google side of Google Glass would then reflect hermeneutic relations. But what about Google Glass itself, as a unified design product?

Using Google Glass as a pair of glasses and for navigational purposes involves both embodiment and hermeneutic relations. The difficulty comes in in that the first involves looking through the Glass, while the second involves looking at it. As has been pointed out with regard to the augmented reality windshield, of which Google Glass can be seen as a "close cousin," it is difficult to see how it would be possible to relate to the same technology simultaneously with both forms of relations without one interfering with the other and so resulting in distracted attention (Michelfelder, 2014). Here is a way that a postphenomenological analysis, by serving as a cautionary note, might be able to play a role in the process of design.

In short, all of the above points to a highly promising future for postphenomenology: a future where key foundational elements of its conceptual scaffolding are further defined, organized, elaborated upon, and modified. The end result is a more insightful understanding of our technological experiences, including how technologies work to shape our individual experiences and social practices. Postphenomenology at the present moment offers up its richest insights when it is in Rosie-the-Riveter mode: rolling up its sleeves, so to speak, in order to work out a close analysis of human-technology relationships, both relations with a single technological device or a handful of technologies which are connected to one another through the experiences of their users. Note I did not said: "human-technology-world relationships." This brings us to the second future I would like to imagine here for postphenomenology.

## POSTPHENOMENOLOGICAL FUTURES: TAKE TWO

Let me begin with a passage with which many readers of this volume will likely be familiar: Section 12 of Martin Heidegger's *Being and Time*. If all the sections of *Being and Time* were ranked according to the quality of philosophical insights they contained, Section 12 would arguably be found toward the bottom of the list. Its purpose is primarily a transitional one: to map the immediate road ahead. Having already identified Dasein as being-in-the-world to be the phenomenon to be questioned, Heidegger pauses in this section to note that the fact that being-in-the-world is a unified phenomenon does not prevent it from being approached from a perspectival point of view. And so, the immediate road ahead will not take the form of a direct philosophical plunge into an analysis of the phenomenon as a whole, but rather will be staged, divided into three parts. First up will be a look at the structure of "in-the-world." Next to come will be a consideration of the "who" that has "being-in-the-world" as its being. Following these two sections, and wrapping up, will be an analysis of "being-in" (Heidegger, 2010, 53–59).

It is in the context of the first of these stages, when Heidegger looks at the structure of "in-the-world," that he lays out what is customarily thought of as his "account of tool use," an account to which postphenomenology owes much in terms of its lineage (Ihde, 2010). But, one could take the section in which Heidegger offers an account of tool use and describe what is going on in it in other ways. This account is, after all, stage-setting. It is a means that will allow Heidegger grounding for his exploration of how the world as a whole becomes disclosed. So, rather than talking about "Heidegger's account of tool use" one could also say: "Heidegger's account of how the world is disclosed." Of course, any particular section in *Being and Time* is preparatory for the next, so one could say that to make this re-description does not cast into any doubt that Heidegger is in this particular section giving an account of tool use. And, indeed, I am not trying to put this account in any doubt. But the fact that Heidegger has his sights set on talking about world in this passage sets the primary focus of its investigation. Another way of putting this would be to say that in Heidegger's exploration of tool use, it is the world more than tools or equipment itself that is at stake.

But, the world has not tended to be foregrounded within postphenomenology as a subject of investigation; even though it is one of the key components of the "I-technology-world" structure that forms the bedrock for postphenomenological analysis. It has, as it were, slipped more into the background of postphenomenological approaches to understanding technology than it has remained out front. Ihde has paid the most attention to it, as seen for example in his perceptive readings not only of Heidegger's exploration of equipment in *Being and Time* ("...what is ultimately revealed is the world as a whole..." [2010, 47]), but in his later writings as well. But I think it is fair to say that overall "world" is the least attended to component in the "I-technology-world" trio in postphenomenological inquiry.

For instance, while I expect he may disagree with me, it seems fair to say that in developing the concept of mediating as co-shaping, Verbeek has not focused extensively on how technology discloses the world as a whole. This is not to say that any discussion of "world" is completely absent from his work. For example, in *Moralizing Technology* (2011, 15) in the context of showing how postphenomenology departs from the modernist perspective on "human-world" relationships; he notes that such relationships:

> should not be seen as relations between pre-existing subjects who perceive and act upon a preexisting world of objects . . . what the world "is" and what subjects "are" arises from the interplay between humans and reality; the world that humans experience is "interpreted reality," and human existence is "situated subjectivity." (2011, 15)

This having been said, however, Verbeek moves on to quickly emphasize that what this means is that "humans and technologies do not have separate existences anymore but help to shape each other in myriad ways" (Verbeek, 2011, 16). This claim then becomes thematic for the remainder of the volume. In the course of drawing out distinctions among different types of rebound effects, he observes that "technology influences human behavior, and, conversely, existing patterns in human behavior influence the use and even the functionality of technologies" (93). A similar observation occurs in his discussion of how mediation functions in the context of ambient and pervasive computing, for which the FoodPhone (a phone that allows users to transmit photographs of the food that they eat and receive an analysis of their calorie content in return) is the salient example. Here, Verbeek identifies the FoodPhone as a form of hermeneutic technology, as it "helps to develop new interpretations of food and consequently informs people's eating practice" as well as social practices associated with eating (126–127). That the FoodPhone, or perhaps even its more recent and more subtle cousin, the Nike Fuelband, mediates both our individual experiences of eating food and our social practices surrounding food can be defended without reference to the world as a whole; put otherwise, to say that technologies disclose patterns of behavior or practices of daily life is not the same as to say they disclose the world as a whole.

Before returning for a second look at behavior-influencing technologies, I want to raise the question as to why the phenomenon of world has not remained more forcefully thematic for postphenomenology. I can think of several, admittedly speculative, reasons why this is the case. One is that the notion of "world" brings to mind other "wholes" such as "utopias" or "dystopias," and it is dystopian thinking in general about technology which postphenomenology attempts to serve as a counterweight. Another has to do with the desire, as Verbeek eloquently puts it, to "do justice to the concrete empirical reality of technology" (2005, 100); this desire leads directly to focusing on specific technologies and their mutual interactions, and so to particular practices and patterns rather than the world as a whole. A third reason is that as a non-foundational philosophical movement, postphenomenology has reasonably been wary of looking into how the world as a whole is disclosed. Its non-foundationalism goes hand-in-hand with the emphasis on particular contexts of human-technological relations; "world," though, particularly in a Heideggerian setting, carries with it a fair amount of ontological baggage. Going back for a moment to Ihde, what has been significant for him with regards to understanding the critical passage about the hammer in Heidegger's discussion of equipment is how Heidegger takes human-technology relations to be relations of embodiment, to the exclusion of other forms of relations. But it is at the point where Heidegger puts his attention on the compromised "workability" of a tool that the world as a whole is disclosed; it is a lack that brings the world as a whole into being, and what might bring the world as a whole into being with other forms of technological relations is a matter remaining for postphenomenological investigation.

Why does it matter though that "world" does not more enter the picture more powerfully for postphenomenology? Why might putting more emphasis on the "world" in the constellation "I-technology-world" be important for its future? The "wholeness" of the world as such can be taken to be a wholeness within which our experience is disclosed to us in a familiar and trustworthy way, a whole in which our interactions with material objects and with others can be counted upon to hang together. For Heidegger, this disclosure, in the context we have been discussing here (Heidegger, 2010, section 17) is non-thematic, amounting to an environment in which we are caught up, in which we can find our whereabouts, to which we are, going back to the expression of Merleau-Ponty's mentioned earlier, "geared-in." If postphenomenology inquiry were more attentive to the forms of disclosure allowing for our "meshing" with the world, it might realize the importance of giving not only more but also critical attention to technologies that threaten to undermine this "gearing-in."

The particular technologies-in-the making I have in mind here, some already deployed and some lying on the horizon, are ones that operate under the threshold of everyday phenomenological experience as this experience is rooted in ordinary perception. The technologies, some already deployed and some lying on the horizon, are ones that look at or gather information from us, particularly from our physical bodies along a variety of dimensions, for the purposes of influencing our future behavior. As indicated earlier, the focal point of much postphenomenological investigation has been on the user experience of individual technologies. Any technology that would work by being hidden from the realm of the everyday phenomenal experience of the user—that would, to put it another way, be "off the grid" from the user experience—could not directly be the focus of a postphenomenological investigation. A different way of putting this would be to say that if we were to talk about how an individual through this technology intends or is directed toward the world in a particular way, it would make no sense. But, such technologies can in fact work to co-shape human behavior in a way that could serve to help diminish overall trust in the world that one experiences.

In short, in this second future postphenomenology would not only be interested in analyzing human technological experience but also, in part, be concerned with safeguarding the disclosive character of everyday phenomenological experience to begin with. In order to get to this point, attention needs to be given to all elements in a matrix of "I—technologies-in-the-making—world," where these technologies are ones that could have the effect of making the lived connection between our experience and the world as a whole more fragile if not to some degree shattered. Such a "speculative postphenomenology" would also have a normative element to it, as it would be interested in maintaining the trustworthiness of the connection between human experience and the world.

## CONCLUSION

Let me wrap up by giving some examples of what I have in mind by technologies that threaten to undermine the disclosive character of everyday phenomenological experience.

In October 2012, at a select number of its locations, the American company Nordstrom began using wi-fi to track its customers. It kept track of how much time its customers spent within its stores in general, and also in particular departments of these stores; it claimed it was not tracking customers' movements from one department to another (Cohan, May 2013). From a technological viewpoint, the set-up by which this tracking took place was fairly simple: sensors placed throughout the store would pick up wi-fi signals sent by customers' smartphones and transmit them to a third-party service which in turn would generate an "online dashboard" available to managers who would use the information gathered to make changes—for example, adding extra staff—where necessary.

Nordstrom's project ended abruptly after customers complained it was invading their privacy. From a postphenomenological point of view, that is not the key issue. The key issue is that the behavior being tracked was disconnected from individual experience in at least two ways. First, because sensors and wi-fi were involved, the customer could not have the phenomenal experience of "being tracked," as she would for example if she were to be watched by video cameras. Second, the behavior being tracked is behavior that normally individuals do not pay attention to themselves. An analogous kind of tracking is used on some shopping websites, where users will receive a pop-up message after a particular (but unknown from the user's end) amount of time has lapsed, pointing out that it looks as though they are having trouble making up their minds, and asking them if they are in need of customer assistance. Certainly someone may give loose attention to how much time they are spending while shopping, but not with regard to how much time they are spending at a particular counter or display area.

More worrisome however from the point of view of weakening the premise of phenomenology that we are "geared in" to the world through our lived experience are technologies that monitor not our lived, embodied selves but rather the "anonymous body," as Merleau-Ponty called it, communicating this information directly to other technologies so that they can make decisions for the user in-between. "Smart" cars are in the works that would use information gathered from biosensors in the driver's seat or seat belt to make decisions about how "fit" the driver is to operate the car, making adjustments as perceived to be necessary, from sounding alerts to the driver to more deliberately turning off the driver's cell phone (White, 2012). "Smart" textiles incorporated into clothing can track galvanic skin response and other biodata in order to monitor a wearer's level of stress and again, make adjustments accordingly, such as to play music or soothing messages—when the stress level of the wearer, which at least in some cases can go undetected by the person wearing the clothing—is determined to be too high (BBC News, 2010). Such kinds of technologies are qualitatively different from the ones

Verbeek discusses such as the FoodPhone (2005); as these rely on user-perceivable inputs leading to user-perceivable outcomes.

How might these technologies-in-the-making work to have an impact on our experience of the world? How might automotive biosensors serve not simply to influence the experience of an individual driver or shape the general practices involved with driving, but also influence the way the world in general is disclosed in the course of everyday life. For example, would they work to reveal the world as a threatening environment, requiring the consistent and constant "sculpting" or "medicalization" of human behavior in order to diminish its threatening [character]? For postphenomenology to get a purchase on these kinds of questions, it would need to think more speculatively about how they might work to shape the world of human experience. The worst case scenario for postphenomenology's future is that while it could continue to offer insights into a range of technologies of everyday life, it would not have much to say, without taking a speculative turn, about "technologies-in-the-making" which aim, through those who write their "scripts," at influencing and controlling human behavior by working behind the scenes.

Graham Harman has pointed out that philosophies tend to have two patterns of movement to them, the first being a movement of expansion, where the principles it has discovered can be extended into "new territory" (2002, 46). Through increased attention to the "world as a whole," I have been suggesting here that postphenomenology could extend itself into the territory of "technologies-in-the-making" particularly the ones I have just mentioned. In this way, postphenomenology could continue to remain on the forefront of philosophies of technology and continue to develop its potential, as Ihde once put it, to offer "a deepening and more complex appreciation of all of the facets of our technologically textured mode of life" (2010, 84).

## REFERENCES

BBC News. (2010). "Smart Clothes Offer Emotional Aid." June 4. http://www.bbc.co.uk/news/10236143

Cohan, P. (2013). "How Nordstrom Uses Wi-Fi to Spy on Shoppers." *Forbes Magazine.* May 9. http://www.forbes.com/sites/petercohan/2013/05/09/how-nordstrom-and-home-depot-use-wifi-to-spy-on-shoppers/

Harman, G. (2002). *Tool-Being: Heidegger and the Metaphysics of Objects*. Chicago: Open Court.

Heidegger, M. (2010/1953). *Being and Time,* trans. Joan Stambaugh. Albany, NY: State University of New York Press.

Ihde, D. (1990). *Technology and the Lifeworld: From Garden to Earth*. Bloomington: Indiana University Press.

Ihde, D. (1993). *Postphenomenology: Essays in the Postmodern Context*. Evanston, IL: Northwestern University Press.

Ihde, D. (2002). *Bodies in Technologies*. Minneapolis, MN: University of Minnesota Press.

Ihde, D. (2010). *Heidegger's Technologies: Postphenomenological Perspectives.* New York: Fordham University Press.

Ihde, D. (2012). *Experimental Phenomenology, Second Edition: Multistabilities.* Albany, NY: State University of New York Press.

Merleau-Ponty, M. (1962). *The Phenomenology of Perception,* trans. C. Smith. London: Routledge & Kegan Paul.

Michelfelder, D. (2014). "Driving While Beagleated." *Techné: Research in Philosophy and Technology* 18(1/2): 117–132.

Rosenberger, R. (2014). "The Phenomenological Case for Stricter Regulation of Cell Phones and Driving." *Techné: Research in Philosophy and Technology* 18 (1/2): 20–47.

Verbeek, P-P. (2005). *What Things Do.* University Park, PA: The Pennsylvania State University Press.

Verbeek, P-P. (2011). *Moralizing Technology: Understanding and Designing the Morality of Things.* Chicago: The University of Chicago Press.

White, J. (2012). "A Car That Takes Your Pulse." *Wall Street Journal.* November 28. http://online.wsj.com/news/articles/SB10001424127887324352004578131083891595840

## *Chapter Sixteen*

# Stability, Instability, and Phenomenology

Albert Borgmann

In the mid-1950s, Kees Boeke, with the help of Els de Bouter, drew a series of forty pictures (Boeke, 1957). The first was the photograph of a girl in a deck chair, holding a cat. She is sitting in the courtyard of the school in Bilthoven, the Netherlands, that Boeke had founded and where he taught. We see the girl from above at a distance of a few meters, say five. In the next picture we have ascended to ten times the distance; so now we see the girl and her surroundings from a point that's fifty meters above her. And so on in jumps of powers of ten up to ten to the twenty-sixth. We then return to the original photograph and move closer to a distance a tenth of the original five meters—fifty centimeters. And so on in jumps of the negative powers of ten, down to the negative thirteenth.

At least four films have been made that have translated the forty jumps into a smooth journey.[1] They start from different places and persons—a man sleeping on a golf course in Miami, a boy with his dog rowing a boat on the Ottawa River, a young couple picnicking on the lake shore in Chicago, children playing on St. Mark's Square in Venice, a woman lying on the lawn of the Googleplex in Silicon Valley. The upward movement of the viewpoint is formalized by changing squares or rings or a bar scale. Sometimes the orbits of the emerging planets are marked out. The final view is of the fairly uniform scatter of galaxies or galaxy clusters.

In all but one of the films the viewpoint returns to the point of departure and then homes in on the person, the skin or the retina of the person, down to a molecule, an atom, and finally down to the atomic particles, stopping in the early films at electrons, protons, and neutrons, and more recently going down to the quarks. These films illustrate and require the analysis of postphenomenology so clearly that they may as well have been titled "Ihde—the Movie." I'll take as my illustration Don Ihde's recent summary of postphenomenology, a statement that sparkles with Ihde's trademarks—energy, accessibility, circumspection, and insight (see 2009). I will follow the four features he stresses there: variational theory, embodiment, lifeworld, and technoscience (Ihde, 2009, 11 and 25–44).

Boeke's book and the films it has spawned illustrate variational theory. "What emerged or 'showed itself'" is the "complicated structure" of the universe (Ihde, 2009, 12). The center of that structure is the lifeworld, the familiar environment of persons, things, and settings. The ascent to the cosmic view and the descent to the microworld depart from the lifeworld and return to it. While variational theory and lifeworld are strikingly illustrated by the book and the films, the themes of technoscience and embodiment alert us to complex assumptions that remain tacit in the presentations of the cosmic journeys. The sweep of first drawing away from the lifeworld and then penetrating it covers up the transition from Ihde's (human-technology) environment relation to the human (technology-world) relation. In the vicinity of the lifeworld, instruments such as ladders, balloons, optical telescopes, glasses, microscopes, and the like enter "into my bodily, actional, perceptual relationship with my environment" (Ihde, 2009, 42). But at certain scales, such tools begin to fail me, and I have resort to "readable technologies" such as the devices that send information from a satellite or are produced by an x-ray machine (Ihde, 2009, 43). Imaging technologies make such information visible and so visualize the invisible (Ihde, 2009, 45–62). The procedures in the book and the films make it look as though vision is homogeneous across these transitions. What gets overlooked is "that somehow along a continuum from our sensory experience phenomena exceeded our bodily capacity to detect" (Ihde, 2009, 54).

Ihde's discussion of how the invisible is rendered visible further reveals that the visual conceit of Boeke's project conceals from view the cosmic information we get from the invisible parts of the electromagnetic spectrum—the x-ray, ultraviolet, and radio frequencies. To this we must add that even all that can be made visible according to Ihde constitutes only a twentieth of all there is. Ninety-five percent of the universe consists of dark energy and dark matter, and what we know of them is only their shadowy dynamic and gravitational effects.

Inevitably, when something is moved to the bright center of attention, something else moves into the shadows as Eugen Fink, Husserl's last assistant, has pointed out.[2] What Ihde illuminates so brilliantly are multistability and variability. It's testimony to the rank of his philosophy that he remains aware and reminds us of what is shaded; and when the multilayered mediations of sensors and digital and imaging technologies threaten to dissolve cosmic reality, Ihde stresses that the opening afforded by these mediations "remains 'realist' if by this is meant that a sensor device only operates if it actually detects some emission" (2009, 61). So what follows is a complement rather than a criticism of Ihde's work.

I want to move the invariant stabilities from the margins to the center. Stabilities should not be taken as necessarily static or abstract. They can be historical and uniquely concrete. They underlie and enable variations and multistabilities. Without some stable and identifiable thing at the center, variants would be different independent entities, and the multistability of interpretations would turn into a multiplicity of objects. Stabilities tend to hide, however, because they are not immediately and fully graspable. What do the following signs refer to?

III<br>
. . .<br>
11<br>
trois<br>
tres<br>
drei

The reference, of course, is three. But, to put it in Fregean terms, "three" does not deliver the reference, but presents just another sense of 3, the sense that is natural to speakers of English. So is "3" for people who are familiar with Arabic numerals and "11" for those conversant with binary notation. In Ihde's "stage/pyramid/robot" example, it seems as though "this configuration, an abstract drawing" captures the thing in itself, innocent of interpretation. But in fact that expression asks us to think of eleven lines as drawn and configured (2009, 12).

That things tend to hide does not mean that their identity is in doubt. The person who says "trois" and the person who says "drei" agree that they are referring to one and the same thing. And that in turn does not mean that there isn't at times a question whether two speakers are in fact referring to the same thing or whether a speaker is referring to anything at all. But such questions are intelligible only on the acknowledgment that normally there are identifiable things and coreferential terms.

Ihde's stress on multistabilities is often effective in disabusing us of the naïve and dogmatic belief that our language captures, in Kantian terms, the thing in itself and that, in possession of the thing, we are uniquely able to judge the values and ways in which the thing appears to others. As important, Ihde also shows how in the case of archery the senses of a reference or the phenomena of a noumenon come to life in different kinds of embodiments and practices.

We can now see how variational theory and the structure of multistability refer to complementary procedures. Variational theory begins with the variety of senses or appearances and points at the invariant reference or thing in itself; multistability centers on reference or the thing in itself and from there explores the variety of senses or phenomena. Variability and multistability are reflected in the variety of cultural communities and technical schools where the senses of things come to life. What tends to disappear into the shadows, however, are not only the things themselves, but also the presence of the encompassing community that identifies and responds to the stability of the thing in itself. Implicitly perhaps, Ihde agrees with that point. For him, the crucial invariant that is identified by variational theory is "the complicated structure of *multistability*" (Ihde, 2009, 12). If it were not for the community of Ihde's readers who know multistability as the stable center of his analyses, postphenomenology and technoscience would fall apart into incomprehensibility.

Ihde is rightly concerned with the variety and instability of the senses of things, and he is the master of their diversity. But there is another concern, perhaps as important, that regards the variety of the things themselves—which ones should be of paramount concern to us? Boeke's book and the subsequent films can be thought of as the various disclosures of two stable structures that inevitably encircle the lifeworld, the outermost horizon of the cosmos and the innermost horizon of the atom. They correspond to two abiding human concerns. Humans, for as long as they have been known to us, have looked up at the sky and situated themselves in an understanding of how the cosmos came to be and how its story and structure informed their lives. Humans have always also looked down to identify the basic stuff the world is made of and how such stuff explains what closely surrounds us and how it can be shaped to our benefit.

In addition to these two horizons, the outermost of the cosmos and the innermost of the atom, I will set aside three more that envelop and inform the lifeworld, two clearly moral ones—global warming and global justice—and one nebulously technological and cultural—cyberspace. How these horizons morally inform the lifeworld is one of the great unanswered questions of contemporary philosophy and requires a new phenomenology.

I mean "moral" in a brought cultural but also distinctive sense, distinctive in the sense that any one horizon should specifically flavor and texture contemporary reality and make our lives more or less worth living. When it comes to the outermost and innermost horizons, that sense, to a first approximation, can be delimited against uncontroversial ways in which our knowledge of the cosmic and atomic structures envelop or inform our lives. Today the cosmic horizon envelops Earth in a spatiotemporally distinctive way. In relation to cosmic space, Earth has been shrinking ever since Copernicus, and the farthest cosmic horizons are bounded by time. Features that haven't had time to reach us are invisible to us. Moreover, our universe may be enveloped by a multiverse that's in principle unobservable. Our knowledge of atoms and their particles has obviously informed and transformed our lives through technology.

There have of course been proposals as to the human condition in the universe. There is the famously pessimistic view of Steven Weinberg that "the more the universe seems comprehensible, the more it also seems pointless" (1988, 154). There is the optimistic view of Joel Primack and Nancy Ellen Abrams that our position at the cosmic center in seven senses of that notion will inspire gratitude and environmental responsibility (2006, 270-272). But existential despair and environmental concern are not specifically or uniquely inspired by cosmic awareness.

These two views have been advanced by physicists. Mainstream philosophers have not developed standard approaches to the relation of the molar to the cosmic world.[3] As regards the relation of the molar to the atomic world, there have been two widely used ways of articulating the problem though there are no widely accepted solutions. Those guiding concepts are supervenience and emergence. We can think of them as looking from the lifeworld down to the atomic world (supervenience) and looking from the atomic world up to the life world (emergence), just as, broadly speaking, multistability looks from a thing to its appearances and variation looks from its appearances to a thing.

Supervenience says that the lifeworld supervenes on the atomic world, and that in turn means that there can be no change in the (supervenient) lifeworld without a change in the (subvenient) atomic world. A green tomato cannot turn red without changes in the atomic or molecular structure of the tomato. The converse is not true. There can be many a change in the molecules of the tomato without it changing its color. Emergence says that when atoms and molecules are compounded into more and more complex structures, properties emerge in the compound that the components lack. Tomatoes have color, but atoms don't.

Supervenience has one important consequence for the molar world or lifeworld. All phenomena are subject to the closure of physics. It cannot be the case that in examining a phenomenon we come upon a substance that is nonphysical or on a causal connection that fails to be physically lawful. For example, there cannot be an immaterial soul substance or a telepathic interaction. But the closure of physics is a very wide constraint on phenomena and gives phenomenology no particular guidance. Like multistability, it takes the existence of things for granted.

Variation and emergence, however, are concerned with the disclosure of things, the invariant bearers of emergent properties; and here again "invariant," like stability earlier, should not be taken as eternal or transcendental, but as variably identifiable here and now. Thing, in Latin, is *res*, and the predicate that goes with res is "is real." Thus attention to the things that lend our lives coherence requires as an addition to postphenomenology a real phenomenology. It's a project entirely in its own right. It, if anything, will answer the question that emergence poses.

"Do Things Speak?" asks Ihde and answers in the affirmative. Again his emphasis is on the variety of languages in which things speak or are allowed to speak. The great little book concludes with the words: "Things, too, have or may be given voices" (Ihde, 2009, 88). To Ihde's question and to his answer the complements are: Which things speak decisively? And how do we learn to listen to them? For Husserl phenomenology was the enterprise that was never finished. Real phenomenology is the enterprise that hasn't really begun.

## NOTES

1. See "Cosmic View," http://en.wikipedia.org/wiki/Cosmic_View, visited March 21, 2014.
2. See Fink, 1957, pp. 321–337, particularly p. 336.
3. One of the exceptions is Tim Maudlin's "The Calibrated Cosmos" (2014).

## REFERENCES

Boeke, K. (1957). *Cosmic View.* New York: John Day Company.

Fink, E. (1957). "Operative Begriffe in Husserls Phänomenologie." *Zeitschrift für philosophische Forschung*, vol. 11.

Ihde, D. (2009). *Postphenomenology and Technoscience: The Peking Lectures.* Albany, NY: SUNY Press.

Maudlin, T. (2013). "The Calibrated Cosmos," http://aeon.co/magazine/nature-and-cosmos/why-does-the-universe-appear-fine-tuned-for-life/, visited March 21, 2014.

Primack, J. R., and N. E. Abrams. (2006). *The View from the Center of the Universe.* New York: Riverhead Books.

Weinberg, S. (1988). *The First Three Minutes.* New York: Basic Books.

# Index

# About the Contributors

## EDITORS

**Robert Rosenberger** is an assistant professor of philosophy at the Georgia Institute of Technology in the School of Public Policy. He studies the philosophy of technology, investigating topics such as Mars imaging, cell-phone induced driving impairment, public space design, neurobiological sample freezing, internet personalization, dictation technologies, and computer-simulated frog dissection. He is the Editor-in-Chief of the new Lexington book series entitled "Postphenomenology and the Philosophy of Technology," which he co-founded with Don Ihde and Peter-Paul Verbeek. He has also edited the 2010 interview book *Philosophy of Science: 5 Questions*, and is the co-curator of the website postphenomenology.org.

**Peter-Paul Verbeek** is professor of philosophy of technology and chair of the Philosophy Department at the University of Twente (www.utwente.nl/gw/wijsb/organization/verbeek/english/index.html). He is also president of the Society for Philosophy and Technology (www.spt.org). His research focuses on the relations between humans and technologies, and the social and cultural implications of technology. Currently he is leading a five-year research project to develop an analysis of the mediating role of technology in knowledge, morality, and metaphysics. Among his publications are *Moralizing Technology: Understanding and Designing the Morality of Things* (University of Chicago Press 2011) and *What Things Do: Philosophical Reflections on Technology, Agency, and Design* (Penn State University Press 2005).

# CONTRIBUTORS

**Kirk M. Besmer** teaches philosophy at Gonzaga University. His research interests include phenomenology and the philosophy of technology. He is especially interested in the ways that contemporary technologies alter our experience and understanding of human embodiment. He is the author of *Merleau-Ponty's Phenomenology: The Problem of Ideal Objects* (Continuum, 2007).

**Albert Borgmann** is Regents Professor of Philosophy at the University of Montana, Missoula, where he has taught since 1970. His special area is the philosophy of society and culture. Among his publications are *Technology and the Character of Contemporary Life* (University of Chicago Press, 1984), *Crossing the Postmodern Divide* (University of Chicago Press, 1992), *Holding On to Reality: The Nature of Information at the Turn of the Millennium* (University of Chicago Press, 1999), *Power Failure: Christianity in the Culture of Technology* (Brazos Press, 2003), and *Real American Ethics* (University of Chicago Press, 2006).

**Frances Bottenberg** currently holds a post-doctoral fellowship in the teaching of Philosophy at Elon University, North Carolina. Her research areas include philosophy of emotion, theories of rationality, and the study of dementia, as well as philosophy of pedagogy. Representative forthcoming publications include "Judging Inappropriateness in Actions Expressing Emotion," "Emotion as the Animation of Value," and a philosophy of education textbook, co-authored with Anthony Weston, titled *How Should We Teach? Eighteen Philosophies of Education In and Beyond the Classroom*. Bottenberg completed her graduate work at Stony Brook University, where she was an active participant in the Technoscience Research Group, lead by Don Ihde.

**Annamaria Carusi** is associate professor in philosophy of medical science and technology, at the University of Copenhagen. She is interested in connections between epistemic, social, and aesthetic aspects of science, with a particular focus on material practices and technologies. On the theoretical level, she works on developing connections between philosophy of science in practice and phenomenology; her research is convergent with new experimentalism and new materialism. Her recent research has focused on images, models, simulations, and visualizations in the contemporary biosciences, and on the role of technologies in computationally intensive interdisciplinary settings.

**Andrew Feenberg** is Canada Research Chair in Philosophy of Technology in the School of Communication, Simon Fraser University, where he directs the Applied Communication and Technology Lab. His recent books include *Between Reason and Experience: Essays in Technology and Modernity* (MIT Press), *(Re)Inventing the Internet* (Sense Publishers), and *The Philosophy of Praxis: Marx, Lukács and the Frankfurt School* (Verso).

**Jan Kyrre Berg Friis** has a PhD in science studies and is an associate professor of philosophy of science and technology at Copenhagen University. He has authored, edited, and co-edited numerous books on philosophy of technology, philosophy of time, metaphysics, and science. He has published papers on topics such as philosophy of medicine, time concepts in physics, time as experience, perception, hermeneutics, and measurement.

**Aud Sissel Hoel** is a professor of media studies and visual culture at the Norwegian University of Science and Technology. Her work focuses on technological mediation, and especially, on the roles of images in knowledge and being. This research interest branches out to include photography, scientific instruments, new media, medical imaging, and visualization. An overarching aim that cuts across Hoel's various projects is to rethink images in dynamic terms, developing what she has coined as a "differential" theory of images and of mediating apparatuses more generally. Her publications cover a wide range of topics on the overlapping fields of visual studies, science studies, philosophy of technology, and media philosophy. For more information, see www.audsisselhoel.com and https://ntnu-no.academia.edu/AudSisselHoel.

**Don Ihde** is Distinguished Professor of Philosophy, Emeritus, Stony Brook University. A founding figure in postphenomenology, he is the author of twenty-one single-author and seven edited or co-edited books. These include: *Postphenomenology: Essays in the Postmodern Context* (1993), *Postphenomenology and Technoscience: The Peking University Lectures* (2009), and, related, Evan Selinger's *Postphenomenology: A Critical Companion to Ihde* (2006). Ihde has lectured in approximately fifty countries, and his works have been translated into fifteen languages.

**Chris Kaposy** is an assistant professor of health care ethics in the Faculty of Medicine at Memorial University in St. John's, Newfoundland, Canada. His main research interests are in bioethics. Current research projects include a study of clinical ethics issues in the treatment of HIV, an empirical project on the roles played by ethics consultants in crisis situations, and the ethics of prenatal testing for Down syndrome. Alongside his research activities, Chris is a clinical ethicist and program director for the Master of Health Ethics degree at Memorial University.

**Asle H. Kiran** is a researcher of ethical, social, and existential consequences of new technologies in the Department of Philosophy and Religious Studies at the Norwegian University of Science and Technology. His research is both empirical and conceptual, and mainly focuses on how new technologies shape and re-shape healthcare practices. Themes of interest include the technological constitution of person and patienthood and the limits of doing proactive responsible innovation. He has published in international and Norwegian journals and books.

**Lenore Langsdorf** is Professor Emerita, Southern Illinois University, Carbondale; visiting research professor, University of Texas, San Antonio, and adjunct instructor, Northwest Vista College. Her research focuses on cognition, the philosophy of the human/social sciences, and the philosophy of communication, approached through hermeneutic phenomenology, postphenomenology, and process pragmatism.

**Diane Michelfelder** is professor of philosophy at Macalester College in the United States. A former president of the Society for Philosophy and Technology, she is currently co-editor-in-chief for *Techné: Research in Philosophy and Technology*, the journal of SPT. Her primary research interests are in the philosophy of technology, especially in ethical questions arising from "smart" technologies and the Internet itself.

**Marie-Christine Nizzi**, a former ENS student, got her PhD in philosophy from the Sorbonne (France). After earning a master's in neuropsychology, she is now working on a PhD in psychology at Harvard, in collaboration with Lille 3 University. Her work focuses on resilience and the sense of self in patients with a history of trauma or neurological disorder.

**Søren Riis** is associate professor of philosophy at Roskilde University, Denmark. He holds a PhD from Albert-Ludwig University, Freiburg (Germany). In his research he focuses on technologies, how they co-shape society and the way we humans think and behave. He has co-authored and co-edited *New Waves in Philosophy of Technology* (2008) and the *Oxford Bibliographies* entry "Philosophy of technology." Recently he published a critical examination and interpretation of Heidegger's understanding of technology "The Ultimate Technology: The End of Technology and the Task of Nature" (2013).

**Adam M. Rosenfeld** currently teaches courses in ethics, epistemology, and philosophy of technology at North Carolina A&T State University. His research focuses on philosophy of science, philosophy of technology, and the philosophy of games, play, and make-believe. A guiding theme through all of his research, that the hermeneutic dimensions of embodied technological mediation are a crucial aspect of all epistemic inquiry, is heavily informed by his participation in the Technoscience Research Group at Stony Brook University.

**Fernando Secomandi is** adjunct professor of product design in the Institute of Arts and Design at Federal University of Juiz de Fora, Brazil. He has published articles on postphenomenology in *Design Issues* and *Design Philosophy Papers*.

**Yoni Van Den Eede** is a postdoctoral fellow of the Research Foundation—Flanders (FWO) affiliated with the Centre for Ethics and Humanism and the Centre for Media Sociology, both at the Vrije Universiteit Brussel (Free University of Brussels). He conducts research into the philosophy of technology, media theory, and media ecology, with an emphasis on phenomenological, cultural, and existential themes. He is the author of *Amor Technologiae: Marshall McLuhan as Philosopher of Technology* (Brussels, VUBPRESS, 2012).

www.ingramcontent.com/pod-product-compliance
Lightning Source LLC
LaVergne TN
LVHW091334280926
848997LV00053B/8